AF772156

Environmental History and Tribals in Modern

Velayutham Saravanan

Environmental History and Tribals in Modern India

palgrave
macmillan

Velayutham Saravanan
Jamia Millia Islamia
New Delhi, India

ISBN 978-981-10-8051-7 ISBN 978-981-10-8052-4 (eBook)
https://doi.org/10.1007/978-981-10-8052-4

Library of Congress Control Number: 2018930663

© The Editor(s) (if applicable) and The Author(s) 2018
This work is subject to copyright. All rights are solely and exclusively licensed by the Publisher, whether the whole or part of the material is concerned, specifically the rights of translation, reprinting, reuse of illustrations, recitation, broadcasting, reproduction on microfilms or in any other physical way, and transmission or information storage and retrieval, electronic adaptation, computer software, or by similar or dissimilar methodology now known or hereafter developed.
The use of general descriptive names, registered names, trademarks, service marks, etc. in this publication does not imply, even in the absence of a specific statement, that such names are exempt from the relevant protective laws and regulations and therefore free for general use.
The publisher, the authors and the editors are safe to assume that the advice and information in this book are believed to be true and accurate at the date of publication. Neither the publisher nor the authors or the editors give a warranty, express or implied, with respect to the material contained herein or for any errors or omissions that may have been made. The publisher remains neutral with regard to jurisdictional claims in published maps and institutional affiliations.

Cover illustration: Inok

Printed on acid-free paper

This Palgrave Macmillan imprint is published by Springer Nature
The registered company is Springer Nature Singapore Pte Ltd.
The registered company address is: 152 Beach Road, #21-01/04 Gateway East, Singapore 189721, Singapore

Preface

This book attempts to analyse the environmental history of India from the period of colonial rule to the early twenty-first century from a historical perspective. Further, it tries to describe the intervention of various factors during the same period.

The main objective of this book is to analyse the unexplored areas of environmental history in the late eighteenth, nineteenth and early twentieth centuries of the colonial period and the second half of the twentieth century and first decade of the twenty-first, that is, during the post-independence period (1792–2007). The focus, however, is largely confined to the Madras Presidency for the colonial period and Tamil Nadu for the post-independence period.

The existing literature on environmental history as a whole is confined to the colonial period, the major themes being either conservation or commercialisation. Both are by and large confined to a particular region or a particular time period. No studies so far have attempted to look at the entire colonial period and almost no studies have been attempted for the post-colonial period. In fact, most environmental history literature is focused on the second half of the nineteenth century. In other words, the late eighteenth, early nineteenth and twentieth centuries of the colonial period and the second half of the twentieth century and the early twenty-first century post-independence have not received adequate attention.

Given this background, this book attempts to analyse the environmental history of south India from the late eighteenth century to the early twenty-first century during both the colonial and post-colonial period. It focuses on five important aspects. (1) How colonial policy impacted on the environment during the pre-Forest Act era and what the major interventions were that impacted on environmental consequences during that period. (2) Whether conservation measures really protected forest resources or continued to commercialise them, and the impacts that they made on people who depended the forest for their livelihood during the post-Forest Act colonial era. (3) After independence, whether the government changed its policies to conserve the forest and to ensure the livelihood of tribals and other forest-dwellers through appropriate policy measures or whether the same colonial policies continued. (4) In the name of conservation, how the state institutions committed atrocities on subalterns, thereby causing a basic violation of fundamental rights in the democratic system. (5) Whether post-colonial environmental policies were really oriented towards environmental protection or whether instead they were concerned with the popular vote and politics. In essence, the book attempts to offer a comprehensive account of environmental history both of the colonial and post-colonial period from the early nineteenth to the early twenty-first century (1792–2007).

It is hoped that this volume will be a valuable contribution to the current discourse on environmental history. Until now, the debate has focused on the later part of the nineteenth century in different regions of India. A comprehensive account of both the colonial and post-colonial period, and how they have either been ignored or not received due attention in the environmental history literature, is presented here, in an attempt to provide in-depth understanding of colonial intervention and environmental history for both the colonial and post-colonial regimes in India.

New Delhi, India Velayutham Saravanan

Acknowledgements

I thank Prof. M. Atchi Reddy, who guided me in exploring the archival materials as I studied the economic history of tribals in Madras Presidency, which helped me to move further towards environmental history in the later period of my career. Indeed, his vast experience and knowledge of archival sources gave me an opportunity to get to the root of various issues and problems from the late eighteenth to the mid-twentieth century. I thank Prof. Mahesh Rangarajan and Prof. K. Sivaramakrishnan for their support and encouragement, which was further rejuvenated as I worked on environmental history literature from the two decades since 1997. In fact, I don't have the words to express my thanks for their valuable suggestions, motivation and immense support. I thank Prof. Michael Fisher and Prof. Douglas E Haynes for their assistance in this endeavour, and I offer my gratitude to Mr P. Shanmugam, General Secretary of the Tamil Nadu Tribal Association, who provided documentary evidence and patiently clarified my thoughts through several discussions.

I thank M.C. Rajan for his critical input and discussion on a wide range of issues, which gave me better understanding of complex issues. I must acknowledge the courtesy that was extended to me by his family over the last three decades. I am also thankful to my friends, particularly M. Sivaraj, P. Periyasamy and the late P. Thangapazham, for their cordial support, and for looking after me during my lengthy stay at Chennai while I was collecting archival data.

My special thanks are due to the staff members of the Tamil Nadu State Archives, who provided easy access to thousands of volumes of archival records over five and half years. My sincere thanks are expressed to the staff of the following libraries: University of Hyderabad, Osmania University, Centre for Economic and Social Studies, Connemara State Central Library, University of Madras, Madras Institute of Development Studies, Tamil Nadu State Assembly Library, and Salem and Dharmapuri Districts Collectorate Libraries.

I must acknowledge and thank a number of editors not only for their encouragement but also for their permission to publish this material. I humbly acknowledge the numerous anonymous referees for their invaluable suggestions that have helped me to enrich the articles I have published on different issues and during different periods of research. I must acknowledge that the first version of several chapters have been previously published: Chap. 2 was published in *International Forestry Review: The International Journal of Forest Science and Policy* (2008); Chap. 3 was published in *Journal of Forest Economics* (2011); Chap. 4 was published in *Modern Asian Studies* (2007); Chap. 5 was published in *The International Journal of Human Rights* (2010); and Chap. 6 appeared in *South Asia Research* (2009). I thank the respective editors of these journals not only for their encouragement but also for granting permission to republish.

I must acknowledge my parents V. Velayutham and V. Chinnammal, who have been a subtle source of inspiration and constant support. My brother V. Ramachandran and his spouse Selvam Ramachandran have encouraged me to succeed further, despite several constraints. While drafting this book, I gained immense support from my spouse Ajitha Saravanan, daughter Soumya Saravanan and sons Gautham Saravanan and Manish Saravanan in completing the task.

Contents

List of Tables

CHAPTER 1

Introduction

European expeditions to explore new geographical terrains, which began in the ninth century and continued until the nineteenth century, primarily to hunt for resources, were inspired by a population explosion, resource saturation and internal conflicts, in addition to European technological advancement. According to Crosby, 'conditions in Europe provided a considerable push-population explosion and a resulting shortage of cultivable land, national rivalries, persecution of minorities—and the application of steam power to ocean and land travel certainly facilitated long distance migration'.[1] This expedition was further supported by climate, wind, immunity and the transfer of plants and animals, adds Crosby. In the early ninth century Europeans moved to nearby islands, particularly to the north-eastern parts of European countries, then later to Africa, America and to South Asia continents, he continues. In other words, Europeans were hunting for new biological resources and in the process several millions settled in different parts of the world. 'Between 1820 and 1930, well over 50 million Europeans migrated to the Neo–European land overseas,' says Crosby.[2] As such, 'the success of European imperialism has a biological, an ecological, component'.[3] Since the beginning of the modern world system, 'the peripheries have been severely exploited for their raw materials and natural products' by the core.[4] Since the early nineteenth century, Britain had exploited forest resources in different parts of the world, such as Ireland, South Africa and some parts of the north-eastern United States,

© The Author(s) 2018

V. Saravanan, *Environmental History and Tribals in Modern India*,
https://doi.org/10.1007/978-981-10-8052-4_1

and it emerged as 'world leader' in the second half of the nineteenth century.[5] The British and other Europeans plundered the resources of new European countries. 'British and European consumers and manufacturers sucked in resources that were gathered, hunted, fished, mined, and farmed in a great profusion of extractive and agrarian systems; sugar from the Caribbean; furs and cod from North America; ivory and cocoa from Africa; spices, cotton, tea and timber from India; wool from the sheep of the Antipodes; rubber from South-East-Asia; gold from South Africa; oil from Middle-East.'[6] In the early phases of imperial expansion, 'commodities that could not be produced in Britain, such as spices, sugar, tobacco, cotton, and tea were particular objects of desire' and in the later phase 'pastoral production such as wool and meat' was the focus.[7] Hence, the main agenda of the European expeditionary movement was to exploit natural resources. In the process, new cropping patterns were developed alongside the exchange of goods. That Europeans started to explore virgin lands to extract natural resources from the ninth to nineteenth centuries is an established fact, and it also turns out to be the origin of environmental history. In other words, the attempt to explore new geographical regions is in itself an inevitable indicator of environmental implications. Beinart and Hughes (2007) point out that 'European imperialism was also inseparable from the history of global environmental change.'[8]

Europeans made contact with the Indian subcontinent from the early sixteenth century and trade and commercial activities continued until the late eighteenth century.[9] The Portuguese, later the Dutch and French and then the English established trade and commercial ventures in the early sixteenth, early seventeenth and late seventeenth centuries respectively. According to Chaudhuri, in the first two decades of the sixteenth century the Portuguese 'policy was aimed systematically at a comprehensive control of the spice trade'.[10] In the early sixteenth century, important export materials were textiles, cotton yarn, cotton fabrics, pepper, rice, pulses, wheat, oil, coconut products, ginger, turmeric, indigo, sandalwood, sappan wood, benzoin, saltpetre, cloves, diamonds, rubies, coral, seed pearls, wax and lac, tutenage (an alloy of copper, zinc and iron), besides horses and elephants.[11] From that time, Europeans slowly started to exercise control over the subcontinent, exploiting the natural resources for their commercial interests. Initially, their interest was confined to spices and some important trees such as teak on the south-west part of the Western Coast. However, within a short time they realised that these natural resources would not last for long, and hence they started paying attention to regenerating these resources, particularly teak, in order to meet future requirements. Hence, the Nilanbur teak plantation was established in 1844 and others followed elsewhere in the country during the late nineteenth century. As Beinart and Hughes (2007) pointed out, 'while natural resources have been intensively exploited, a related process, the rise of conservationist practices and ideas, was also deeply rooted in imperial history'.[12] While discussing the importance of conservation, it must be understood that the 'growing conservationist concerns did not in themselves stop exploitation'.[13] The British forest conservation idea was not really aimed at protecting jungles and wildlife but at meeting the future needs of the empire. According to Swami, 'Practical considerations such as the need to provide timber for railway and ship-building activities were very important motives for setting up forest conservatories in different parts of India.'[14] In other words, the real intention behind colonial conservation measures was commercial interest, and forest management meant meeting future demand.

The emergence of environmental history began with the study of rational conservation, the use of resources and the preservation of natural spaces that emerged in the United States during the last quarter of the twentieth century.[15] Environmental history largely deals with three broad themes: material, cultural/intellectual and political. The first focuses on the changes in biological and physical environments and how they affect human societies. The second is about representations and images of nature in the arts and letters, and how they have impacted upon society and nature. The third dwells on how law and state policy have changed nature and society.[16] The early literature on environmental history has a consensus that pre-conquest and pre-colonial societies had ecological harmony.[17] 'Most past societies, for instance, have not altered the natural world at anything like the rate or scale that has typified the modern era.'[18]

India's environmental history has largely focused on colonial forestry and the debate centred on the 'contest between commercial forestry and the livelihood and rights of the rural poor'.[19] According to Sivaramakrishnan, 'In much of the existing work on India's colonial environmental history, concerns about how transnational flows of ideas and trained personnel generated a cosmopolitan, professionalised, international environmentalism have been juxtaposed with regional and local political struggles over forests and the lives associated with them.'[20] As such, the debate is focused mainly on two broad themes: pre-colonial equilibrium between nature and society and the commercialisation or conservation of nature and society during the colonial period. The first school of thought argued that prior to colonial intervention an equilibrium between people and nature was maintained and that was disturbed with the advent of colonialism.[21] During the pre-colonial period, an approximate equilibrium of natural resource base existed between 800 and 1800.[22] Shiva also stresses this point: 'forest had been controlled and used collectively by village communities thus ensuring a sustainable use of these renewable resources'.[23] Mahesh Rangarajan puts it aptly: 'The pioneers of environmental history have tended to set up an opposition between the equilibrium between people and nature before colonialism and the disharmony that arose as a result of British intrusion. Previous rulers had rarely intervened in woods and pastures, mainly to assert monopolies over valuable animals like elephants.'[24] Similar arguments have emerged in different regions.[25] Sivaramakrishnan points out that 'The bold, and in places, admittedly controversial grand narrative that people still use as a point of departure for detailed, if contrapuntal studies, had not yet been published.'[26] Hence, it becomes evident that literature on the environmental history of the subcontinent has been largely centred on the colonial impact.

All the same, the second school of thought denied any equilibrium between nature and people from the third century onwards, stating that conflicts persisted in different parts of the subcontinent and that forest resources were exploited for commercial interests.[27] Yet it admits that there was no sharp conflict among the communities over natural resources. Swami pointed out that 'there is as yet no evidence of total collapse or sharp conflicts among communities over ecological issues in the precolonial period like those one can find during the colonial period'.[28] Sivaramakrishnan argues that 'It is fair to say, therefore, that new scholarship from the mid-1990s onwards, especially when drawing empirical evidence from the great river valleys and central Indian or south Indian plateaus, finds it harder to sustain any simple arguments about pre-colonial equilibrium between natural and social systems'.[29]

As far as the second theme, the commercialisation or conservation of nature and society during the colonial period, is concerned, there emerged two sets of arguments. According to Grove, the importance of protecting forests was realised by the colonial administrators during the early nineteenth century at the highest levels, and this was the fountainhead of environmental protection not only in India but also in other colonies.[30] Debunking Grove, scholars point out that there was no effort made towards conservation at this time as commercial plunder was the prime agenda.[31] No attention was given to the adverse impact of this on people's livelihood options. However, in recent times these issues have been brought out on a regional level.[32] Forest conservation initiatives or scientific forest management, which took shape from the second half of the nineteenth century, were confined to the induction of formally trained personnel from the German Forest College into the Forest Department in order to develop scientific forest management.[33] While analysing the emergence of scientific forest management, scholars have not paid due attention to the consequences on forest-dwelling communities' livelihood options or how forest resources were systematically exploited for the commercial interests of the British until the end of the colonial era. Mahesh Rangarajan says: 'Environmental histories of the colonial period are in large part about the fates and fortunes, the strategies and struggles, and the changing practices and consciousness of underprivileged or marginal groups.'[34]

As indicated earlier, the two broad themes of environmental history, pre-colonial equilibrium between nature and society and commercialisation or conservation during the colonial period, have serious gaps in terms of argument and engagement. Of course, they vary from region to region depending upon the proportion of natural resources, density of population and needs of the community. For instance, in the Madras Presidency, most of the hill tribals lived in isolation, marked by a subsistence economy. Hence, any threat to equilibrium and harmony was not made at least until the early nineteenth century. A significant issue, often overlooked in environmental history literature, is that extracting natural resources as a source of livelihood within given boundaries is something different from exploiting natural resources for commercial ventures by external forces. During the pre-colonial period, extracting natural resources was either for livelihood or to expand cultivation, whereas the colonial regime systematically plundered resources while imposing restrictions on the people. Therefore, comparing nature and its utility during the pre-colonial period with the situation under the colonial regime is inappropriate. However, it may be appropriate to study a particular region to discover how nature was managed during the pre-colonial period and the factors that impinged upon natural resources in terms of population, consumption pattern, trade and commercial activities, how those factors changed over the period and how colonialism altered the rules of the game.

Hypothetically speaking, if pre-colonial society had been exploitative in nature for a variety of purposes, including cultivation, trade and commerce, at the time of colonial intervention we would not have had any natural resources today. Hence, comparing pre-colonial natural resources management with colonial commercial plunder is not be an appropriate argument. For instance, until colonial intervention took place teak was to be found in large quantities in the southern parts of Western Ghats and was used by the people for a variety of activities, but it was completely lost within two to three decades of the arrival of the East India Company. Likewise, sandalwood was found largely in the Eastern Ghats but was systematically exploited and exported to European countries. Therefore, livelihood options within the given boundaries could not be compared with exploiting forests and natural resources for commercial interests by external forces. As indicated earlier, nature and society during the pre-colonial period was different from during the colonial era, but the argument has remained largely centred on the same themes over the last three decades. While debating the colonial environmental history of India, it was pointed out that there was not an equilibrium between nature and society during the pre-colonial period but that there were conflicts at different times and in different regions.

More recently, scholars have maintained that the continued exploitation of forests even post–independence takes the sheen off the argument that postulates equilibrium during the pre-colonial period and exploitation during the colonial period, even to the extent that it is invalid, 'contradictory and inconsistent'.[35] Beinart and Hughes argue that 'equilibrium will ever be found, not that society can achieve the elusive goal of sustainability'.[36] Further, 'pressures on the forest did not end with independence'. The current rate of deforestation is said to be well over 1 million hectares every year. Now as previously, 'critics of state policy point the finger at "contradictory and inconsistent" forest policies that ostensibly aim to protect the environment while also milking it for profit'.[37] Hence, the debate on colonial environmental history either takes the pre-colonial period or the post-independence period to substantiate the colonial commercial exploitation of nature. Again, during the modernisation process in the post-independence period, forest resources have been used to lay the foundation for developmental projects. Though the post-independent dispensation did not set an agenda to export forest resources, they were nevertheless utilised for development.

The emphasis on conservation/environmental protection by the empire from the early nineteenth century saw the establishment of a Forest Department, enactment of a Forest Act and scientific forestry management initiatives from the late nineteenth century onwards. Further, it has been argued that forest resources were extracted not necessarily for the empire project even during the pre-colonial period. These arguments have been made at different times. For instance, some scholars have been bowled over by the scientific forest management initiatives, the emergence of forest colleges at a global level and how those they trained were employed in the colonies, planning scientific forest management from the late nineteenth to the mid-twentieth century. The initiatives are largely based at presidency or country level. Against this backdrop, the last quarter of the nineteenth century witnessed the enactment of the Indian Forest Act 1878.

One of the important gaps in the debate is the failure of those who are arguing for scientific forest management to address the issues of forest-dwelling communities whose livelihood options have been curtailed so that unfettered exploitation of forest resources can take place. They have failed to capture how this impacts on the society and environment of a region in a larger perspective. Apart from livelihood restrictions, other factors that have an impact on the environment are population, trade and commerce, migration, development policies, environmental policies, political economy, politics, law and so on. Yet how these came to have a bearing upon nature and society has not gained focus in the debate. As such, the debate has remained confined to forest, pasture and grazing from the late nineteenth to the early twentieth century. Mahesh Rangarajan suggests that the debate should not be confined to pre-colonial versus colonial ecological destruction but should rather focus more on the social and political process that influenced environmental history.[38] For Rovers (2000), 'There is a possible arena of inquiry for environmental historians, or at least an aspect to our inquiries, which involves the recent history of environmental policy. This is needed and relevant, and could be a valuable dimension of environmental history.'[39] Stressing the importance of law and policy, McNeill says that 'law and policy have always had more to say about the land than the sea'.[40]

On a global level, macro-level factors such as population, trade and commerce were taken into account in the process of reconstructing environmental history. However, in India debate has undermined factors such as population, policies on trade and commerce, and infrastructure development, which should have been accorded due attention. Does the intervention of empire and its initiatives have any continuity? Even as conservation was being emphasised, populist vote bank politics played havoc in depleting the environment. Hence, Swami pointed out that 'Much meticulous research needs to be done in order to present the problem of periodising ecological history in India in all its complexity.'[41] A comprehensive understanding of India's environmental history requires an appraisal of all six phases: pre-colonial (1790s), pre-Forest Act (1790s–1870s), post-Forest Act (1870s–1940s), post-independence (until the 1980s), post-Forest Conservation Act (until 2006) and vote bank politics era (2006 onwards).

It is a well-known fact that, during the pre-colonial period and even in the early colonial period, some kind of equilibrium was maintained between society and natural resources owing to the low density of population. Of course, in a few regions various conflicts emerged but they were not prolonged. Although forest resources were exported to other countries, this was very limited. Hence, there were vast tracts of forests and other natural resources in abundance at the time of colonial invasion.

The second phase witnessed a major intervention, since the very agenda of colonialism was to explore natural resources across continents. The European exploration, which commenced in the early ninth century, finally reached Indian shores during the early sixteenth century. That their arrival and taking control resulted in a major threat to natural resources and the ecosystem of the subcontinent is an established fact. From the beginning, they have attempted to loot natural resources in a much more systematic manner and exported them to European countries. In due course, they have taken over ownership rights for a wide range of forest products. The office of the Conservator of Forests was created in 1806 primarily to assess the resources that could be extracted. Since the second decade of the nineteenth century, tea and coffee plantations had been established in different parts of the country. In the third decade, the enumeration of sandalwood trees was carried out so they could be cut in the near future. While doing this, the East India Company realised that the depletion of teak trees meant that regeneration was required to meet the growing demand. This prompted their establishment of teak plantations, starting with one in Nilambur in 1844. Rather surprisingly, it has to be pointed out that there was no major restriction imposed on local communities at that time to access the forest resources!

In the context of introducing scientific forest management and accompanying restrictions on tribals and forest-dwellers, the third phase is very important in the wake of the Indian Forest Act 1878 coming into effect. Interestingly, the debate in Indian environmental history was centred on this phase. The core argument was about conservation versus commercialisation. One argument is that the colonial administration was instrumental in developing scientific forestry with the introduction of trained forest officers from Germany. According to this view, the forest administrative system was set up and strengthened through a formal interactive mechanism and intellectual exercises, thereby bringing a very large proportion of forests under forest reserves. The second argument contends that the process of establishing scientific forest management curbs were placed on the tribals, which resulted in militant protests and rebellion against the colonial administration. This phase witnessed a phenomenal increase in the export of forest wealth, and received huge attention.

The fourth phase is a continuation of the colonial approach towards forests and the forest-dwelling communities. Despite several provisions in the Indian Constitution to protect the tribals, this approach continued until the introduction of the Forest Conservation Act 1980. Various developmental projects annually gobbled up 1.5 lakh hectares of forest land during this period. Development initiatives were made to enhance the livelihood options of the tribals especially after the fifth five-year plan, but ironically they only facilitated the non-tribals. It should be noted, however, that the Forest Conservation Act imposed restrictions on diverting forest land for any projects without central government clearance. Though infrastructural facilities have been built in the hills and forests, non-tribal settlements were not part of them. Therefore, wherever infrastructure was developed forest resources were extracted. In enhancing the livelihood options, populist vote bank politics played a pivotal role that continued until 2006. If we look at the vote share of various parties from the tribal areas since the independence, we can see that the Congress cornered a lion's share and continued its hold for two decades after independence, but declined sharply in the 1970s. To retain the vote bank, a tribal sub-plan concept was introduced in the 1970s. Though this paid rich dividends initially, the vote share declined from the late 1990s and touched rock bottom in the 2004 elections. Consequently, the Scheduled Tribes and Other Forest Dwelling Communities (Restoration of Forest Rights) Act 2006 was enacted to regain the vote bank.

Hence, Indian environmental history cannot be seen as a unique phenomenon that continued for more than two and half centuries. The different phases of intervention have to be looked at in a systematic manner to capture their change, continuity and impact over time. This book therefore examines all these phases and their consequences from the perspective of the environmental history of the southern peninsula of the subcontinent from the late eighteenth to the early twenty-first century.

Existing literature on the environmental history of India is largely confined to the colonial period, and the debate is centred on conservation versus commercialisation. Further, the available corpus is by and large focused on a particular region or a particular period. In fact, most of the research is devoted to the second half of the nineteenth century. It becomes clear that the post-independence period, 1947–2007, has not received requisite attention. Given the background, this book attempts to rectify this.

The environmental history literature of India is largely confined to the colonial period, more precisely the third quarter of the nineteenth century, being centred on conservation or commercialisation. Further, these studies have been confined either to a particular region or to a particular time period, mostly the second half of the nineteenth century, to some extent the early twentieth century. However, research so far has clearly shown that environmental protection measures have adversely affected the sustainable livelihood of the forest-dwelling and forest-dependent communities. Unfortunately, very few studies have focused on the early nineteenth and early twentieth centuries, while almost none is available for the post-independence period. As such, this book attempts to give a comprehensive picture of the nature and trend of India's environmental history, encompassing the colonial and post-colonial period, spread over more than two centuries from the late eighteenth to early twenty-first centuries (1792–2007).

The book focuses on five important aspects. (1) How the colonial policy impacted the environment during the pre-Forest Act era, and what the major interventions were that proved detrimental for the environment. (2) Whether the measures touted to give impetus to conservation have really protected the forest resources or failed to contain commercialisation, and what impact they made on the people who were dependent upon the forests for their livelihood. (3) After independence, whether governments have changed policies to conserve forests and ensure the livelihood of tribals and forest-dwelling communities through appropriate policy interventions or whether they have continued with the same old colonial policies. (4) Whether the post-colonial environmental policies were really oriented towards protection of resources or were dictated in order to appease popular vote-bank politics. (5) How the state institutions committed atrocities on subalterns in the name of conservation, violating their fundamental rights in the democratic system. Precisely, the book attempts to give a comprehensive account of environmental history covering both the colonial and post-colonial period of India, from the late eighteenth century to the early twenty-first century.

Synoptic View of the Chapters

The main objective of this book is to study the environmental history of India from the late eighteenth to early twenty-first centuries. This chapter, the first of seven, attempts to situate the research argument in the debate within environmental history, and show how the issues and problems need to be addressed. Further, it gives a synoptic view of the remaining chapters.

The focal point of Chap. 2 is a critique of colonialism and its projects regarding the economic exploitation of forest resources and the consequences of this for the environment and for hill inhabitants of India during the pre-Forest Act colonial era in a historical perspective (1792–1882). It argues that the focus of the colonial forest policy was predominantly on the extraction of forest resources for commercial purposes, which had a disastrous impact on the environment and the tribal populace. Further, it analyses the colonial policies to show that they were not aimed at conservation but rather at rampant commercialisation. Besides encouraging the commercialisation of forest resources and non-tribal settlements, this has led to the disintegration of the traditional tribal system as well as the forest-oriented economy. Precisely, the British colonial forest policy wilfully refused to accord due importance to the environment.

In Chap. 3 the colonial forest policy and its impact on the environment and tribals in the Madras Presidency during the post-Forest Act period (1882–1947) are analysed. During this period, the colonial regime actively encouraged commercialisation while several restrictions were placed upon the tribals and other forest users. Here it is argued that the initiatives of the state towards conservation were primarily intended to curtail the access enjoyed by tribals and other forest users to facilitate commercial exploitation. The plunder of forest wealth continued unhindered, leading to the denudation of vast stretches of green cover and a drastic decline in the living conditions of tribals in the Madras Presidency.

In Chap. 4 the forest policies of the state government and their impact on the environment and tribals in Tamil Nadu during the second half of the twentieth century (1950–2000) are discussed. It also analyses the contradictory policies framed by the state to conserve forest resources. On the one hand, the several forest acts and rules have been enacted and restrictions have been imposed on the tribals and other forest-dwellers, whose economy is associated with the forests; on the other hand, forest land has been diverted for various development activities, encouraging commercialisation. Given the curbs on the tribals in accessing forest resources, the impact of development initiatives—protective and promotional measures—on their economy is also captured. Further, the state, even while emphasising environmental protection, has encroached upon the forests for various developmental purposes, and the resultant consequences include displacement of tribals. The chapter also tries to capture the role of different departments of the state administration in protecting forest resources and also the nexus between state administration and politicians/contractors in the plunder of forest resources. In short, it will contribute immensely towards understanding government policies and their implications for environment and tribals in Tamil Nadu during the second half of the twentieth century.

The politics of tribal development are dealt with in Chap. 5, which analyses how politics has played havoc with tribals and the environment during the post-independence period. Even after fifty years of Indian planning, the deprived groups, particularly the tribals, continue to remain underdeveloped, with their living conditions further deteriorating. From a historical perspective, this chapter seeks to capture the political undercurrents of economic policymaking towards tribals and tribal development planning and specifically analyses the consequences of the Scheduled Tribes and Other Traditional Forest Dwellers (Recognition of Forest Rights) Act of 2006. Examining whether this new and politically motivated law will provide an impetus for the economic progress of tribals and environmental sustainability, it is argued that it will neither benefit the tribal communities nor enhance conservation. Rather, it serves as a cloak to justify non-tribal intervention, with potentially disastrous consequences.

In Chap. 6 an attempt is made to situate the nexus between different departments of the state administration, forest, police and revenue, in the violation of human rights of the tribal communities in Dharmapuri district of Tamil Nadu during the last decade of the twentieth century (1990–2000). Further, it looks into the unholy alliance between the Forest Department and politicians in looting forest wealth. Despite the plethora of rules, several forest acts and restrictions on the tribals and other forest-dwellers, the functioning of the Forests Department has paved the way for powerful politicians/contractors to plunder the forests. In spite of the official clamour for conservation and protection of the green cover, the tribals' basic human rights were violated by different departments of the state administration. As a focus of the chapter, it is shown how a fabricated incident was used so that an attack could be carried out by the police on hapless tribals, and that a tortuous judicial process ensued in order to secure justice. The role of the Left parties and activists who stood by the poor tribals is also brought into perspective. Concluding observations are provided in Chap. 7, which also shows how the present work contributes to the ongoing debate on environmental history.

Notes

1. Alfred W. Crosby, *Ecological Imperialism: The Biological Expansion of Europe, 900–1900*, Cambridge: Cambridge University Press, 1986, p. 5.
2. Ibid.
3. Ibid., p. 7.
4. Vandana Swami, 'Environmental History and British Colonialism in India: A Prime Political Agenda', *CR: The New Centennial Review*, 2003, 3(3), 113.
5. Ramachandra Guha and Madav Gadgil, *This Fissured Land: An Ecological History of South Asia*, New Delhi: Oxford University Press, 1997, p. 119.
6. William Beinart and Lotte Hughes, *Environment and Empire*, Oxford: Oxford University Press, 2007, p. 2.
7. Ibid., p. 10.
8. Ibid., p. 1.
9. Sanjay Subrahmanyam, *Political Economy of Commerce Southern India 1500–1650*, Cambridge: Cambridge University Press, 1990.
10. K. N. Chaudhuri, 'Foreign Trade', in Tapan Raychaudhuri and Irfan Habib (eds), *The Cambridge Economic History of India*, Vol. 1, Cambridge: Cambridge University Press, 1982, p. 382.

11. Chaudhuri, 'Foreign Trade', pp. 382–433; Subrahmanyam, *Political Economy of Commerce*, Chap. 4.
12. Beinart and Hughes, *Environment and Empire*, p. 1.
13. Ibid., p. 18.
14. Swami, 'Environmental History and British Colonialism in India', p. 119.
15. Fabien Locher, Grégory Quenet, 'Environmental History: The Origins, Stakes, and Perspectives of a New Site for Research', *Revue d'histoire moderne et contemporaine*, 2009/4, No. 56–4, 2009, pp. 7–38 (Fabien Locher, Grégory Quenet, "L'histoire environnementale: origines, enjeux et perspectives d'un nouveau chantier", *Revue d'histoire moderne et contemporaine* 2009/4 (No. 56–4), pp. 7–38. (http://www.cairn-int.info/focus-E_RHMC_564_0007--environmental-history.htm. [Accessed 3 May 2016]), p. 15.
16. J. R. McNeill, 'Observations on the Nature and Culture of Environmental History', *History and Theory*, 2003, 42(4): 6.
17. Ibid., p. 21.
18. William Cronon, 'The Uses of Environmental History', *Environmental History Review*, 1993, 17(3): 14.
19. Mahesh Rangarajan, 'Environmental Histories of India: Of States, Landscapes, and Ecologies', in Edmund Burke III and Kenneth Pomeranz (eds), *The Environment and World History*, Berkeley: University of California Press, 2009, p. 231.
20. K. Sivaramakrishnan, 'Science, Environment and Empire History: Comparative Perspectives from Forests in Colonial India', *Environment and History*, 2008, 14(1): 42.
21. M. Gadgil, and R. Guha, *This fissured land*. Delhi: Oxford University Press, 1992; V. Shiva, *Ecology and the politics of survival: conflicts over natural resources in India*. New Delhi: Sage Publications, 1991.
22. Madhav Gadgil, 'Towards an Ecological History of India', *Economic and Political Weekly*, 1985, 20(45, 46 and 47): 1913.
23. Shiva, *Ecology and the Politics of Survival*, p. 14.
24. Mahesh Rangarajan, 'Environmental Histories of South Asia: A Review Essay', *Environment and History*, 1996, 2(2): 130.
25. For comprehensive review see Velayutham Saravanan (2017), *Colonialism, Environment and Tribals in South India, 1792–1947*, Introduction.
26. K. Sivaramakrishnan, 'Forests and the environmental history of modern India', *Journal of Peasant Studies*, 2009, 36(2): 303.
27. Richard H Grove, *Green Imperialism: colonial expansion, tropical Edens, and the origins of environmentalism, 1600–1860*, Delhi: Oxford University Press, 1995; Sumit Guha 'Forest politics and Agrarian empires: The Kandesh Bhills, c.1700–1850', *Indian Economic and Social History Review*, 1996, 33(2): 133–153; Sumit Guha, *Environment and Ethnicity in India,*

1200–1991, Cambridge: Cambridge University Press, 1999, Chap. 5; David Baker, 'State polity, the market economy, and Tribal decline: the central provinces, 1861–1920', *Indian Economic and Social History Review*, 1991, 28(4): 341–370. See also, Swami, 'Environmental History and British Colonialism in India', pp. 116–118.

28. Swami, 'Environmental History and British Colonialism in India', p. 118.
29. Sivaramakrishnan, 'Forests and the environmental history of modern India', p. 306.
30. Grove, *Green Imperialism*, pp. 385–473; Beinart and Hughes, *Environment and Empire*, p. 111.
31. S. Ravi Rajan, *Modernizing nature: forest and imperial eco-development 1800–1950*, New Delhi: Orient Longman, 2008, p. 55; Gregory Allen Barton, *Empire Forestry and the Origins of Environmentalism*, Cambridge: Cambridge University Press, 2002, p. 44; for details see, Velayutham Saravanan (2017) *Colonialism, Environment and Tribals in South India, 1792–1947*, London: Routledge.
32. See Velayutham Saravanan, *Colonialism, Environment and Tribals in South India, 1792–1947.*
33. Grove, *Green Imperialism*; Rajan, *Modernizing Nature*; Barton, *Empire Forestry and the Origins of Environmentalism*; Beinart and Hughes, *Environment and Empire.*
34. Mahesh Rangarajan, 'Environmental Histories of India: Of States, Landscapes, and Ecologies', in Edmund Burke III and Kenneth Pomeranz (eds), *The Environment and World History*, Berkeley: University of California Press, 2009, pp. 237–238.
35. Beinart and Hughes, *Environment and Empire*, p. 129.
36. Ibid., p. 21.
37. Ibid., p. 129.
38. Rangarajan, 'Environmental Histories of India', p. 238.
39. Stephen R. Rovers, 'On the Contribution of Environmental History to Current Debate and Policy', *Environment and History*, 2000, 6(2): 145–146.
40. McNeill, 'Observations on the Nature and Culture', p. 42.
41. Swami, 'Environmental History and British Colonialism in India', p. 118.

CHAPTER 2

Exploitation of Forests, 1793–1882

Debate on the environmental history of India, which emerged during the last quarter of the twentieth century and predominantly focused on forestry, can be broadly classified into two categories: the commercialisation of forests and their conservation.[1] The first school of thought argues that forest resources were exploited for commercial purposes and that the forest-dwellers and users were neglected, which led to the decline of their livelihood.[2] The other school of thought argues that although forests were commercialised during the early nineteenth century, the concept of conservation was accorded priority during the early colonial era.[3] It argues that the forest-dwellers and users were responsible for deforestation; further, that the commercialisation of forests was not an early nineteenth-century phenomenon but was widely prevalent even before the colonial intervention.[4] Interestingly, the first school's point of argument dates from the late nineteenth century and the second school's argument dates from the early nineteenth century. However, in recent years, scholars have explored the environmental history of the early nineteenth century in more depth and have concluded that the conservation of forests was not given primacy while the state continued to encourage and patronise the commercial exploitation of forest wealth.[5] Pathak further substantiated this. Although conservation initiatives took place in Malabar and Kanara

An earlier version of this chapter was published in *International Forestry Review: The International Journal of Forest Science and Policy*, 10(1), 2008.

© The Author(s) 2018

V. Saravanan, *Environmental History and Tribals in Modern India*,
https://doi.org/10.1007/978-981-10-8052-4_2

regions of the Madras Presidency during the early nineteenth century, the prime concern was to meet timber requirements and revenue demand.[6] The first school of thought's argument is based on the northern parts of the Indian subcontinent, whereas the second school's argument concentrates mainly on the southern parts of the Indian peninsula.

Further, existing tribal studies were devoted mostly to anthropological aspects and concentrated on the small tribal communities of the Nilgiris.[7] Notwithstanding the growing literature on tribal economy, recent specific studies are by and large concerned with the impact of welfare measures, particularly after independence.[8] In fact, no analysis of how the tribal economy was disturbed during the colonial period has been attempted with a historical perspective, especially in south India. Hence, an attempt is made here to analyse the exploitation of forest resources in south India during the pre-Forest Act colonial era, to fill the gap in environmental history's literature.

The main objective of this chapter is to analyse colonialism and its project regarding the economic exploitation of forest resources, to meet the burgeoning demands of the empire, trade and industry, and the consequences of this for forest-dwellers/tribals and the environment in south India during the pre-Forest Act colonial regime (1793–1882). An important research problem here is to ascertain how colonialism perceived the forests and its consequences on the environment and tribals during this period. In other words, it is to understand whether colonialism encouraged the conservation or the commercialisation of forests during the study period. Other related aspects are whether colonial policies encouraged commercial crops, which gradually led to large-scale deforestation, and how they altered the tribals' living conditions and land use patterns, as well as the environment more widely. Asking these questions will be helpful in understanding the colonial policies relating to forests and their consequences on the tribals and the environment at this time .

Forest Resources in South India (up to 1793)

This section reviews how forest resources were used and managed both by tribals and others in south India until the late eighteenth century, prior to colonial intervention. Scholars have written extensively about the tribal livelihood before the intervention of external/market forces in various places in the Madras Presidency.[9] However, giving a brief account of the tribal livelihood and the resource management pattern during the pre-colonial period is worthwhile. Invariably, throughout the Indian subcontinent, the tribals' livelihood was associated with forests and forest resources. The forest provided them with food, housing materials and implements requisite for their survival. In fact, until the external intervention, the tribals' productive activity was 'human interaction with nature'.[10]

The tribals depended solely on the forest and forest resources for their livelihood, managed by themselves and without any restrictions. Despite this, the use of forest resources was subject to three constraints: the size of the community, the needs of the community and the willingness of its members to exert them.[11] Tribal population density was very low: for instance, in the Kolli hills in the Sangam period,[12] it was around 500 people inhabited for the entire hills (170 square miles) during the tenth century.[13] Even at the time of colonial intervention, the population density in the hill areas was very low when compared with the plains. For instance, in the Salem and Baramahal region, where a large tribal population inhabited the hills, the density per square mile was only 174 persons, and this varied from three to eighty-eight persons in different hills at the end of the eighteenth century. This indicates that much of the land in hill regions was left unoccupied at the time of colonial intervention and the tribals had vast tracts of forest at their disposal as a common property resource. In addition to the low density of population, the tribals' needs were limited and they remained at an irreducible minimum owing to their primitive cultivation techniques and limited control over the forces of nature.[14] In other words, because of their simple technology, lack of specialisation and the restriction of individual productive capacities, their livelihood was maintained at subsistence level without much need or scope for exchange.

The earlier tribal economy was marked by the absence of any desire to make profit, either from production or through exchange activities.[15] In fact, the economy was non-market or subsistence. There was also no motive for gain, labouring for remuneration or any separate and distinct interests based on economic motives.[16] Production was mainly for domestic consumption, with informal exchanges within the system to satisfy their needs.[17] No household was in the normal course of events excluded from direct access to the means of its own survival.[18] The tribals treated the hills' natural resources as common property resources.

Their subsistence was drawn from the collection of forest produce and traditional methods of agriculture. They practised two types of cultivation—shifting cultivation and settled agriculture. Shifting cultivation was known as ***kothukadu*** in Tamil.[19] Trees were felled in a selected place and the remaining shrubs were burnt prior to the sowing of seeds with hoes and spades.[20] The consequences of shifting cultivation largely depended on the overall density of population. Recent studies have proved that shifting cultivation, practised in different tribal areas over the centuries, did not affect the sustainability of forest resources.[21] The British policy of restricting shifting cultivation not only brought forests under State control to generate more revenue but also facilitated the commercialisation of the forests.[22] After the Company Raj, the direct rule of the British Crown restricted and discouraged shifting cultivation to serve the cause of empire.[23] It should be remembered that a low population density did not cause any damage to the hill ecosystem.[24] As such, the tribals could not be held responsible for deforestation during the pre-colonial period. One can ask how the impact of shifting cultivation differed in high and low rainfall areas. Numerous studies have demonstrated that in the high rainfall areas shifting cultivation ensured the forest resources' sustainability. It basically depended on seasonal rainfall. In the poor or low rainfall areas, there was obviously little scope to carry out shifting cultivation. Therefore, where there was a low population density and low rainfall, the possibility of practising shifting cultivation was limited. If it did take place, it would have very limited impact on forest resources. It may also be observed that in the Madras Presidency the forest was not densely populated until the nineteenth century. Shifting cultivation, which was in vogue among the tribals, was almost non-existent in the southern parts of the Presidency.[25] One should ask whether livestock would have destroyed the forest through grazing, and whether this could have caused deforestation; but it is clear that population density was very low and the number of livestock that could have been reared and how many trees were needed to meet the demands of huts and fuel-wood were low. Therefore, the closed subsistence economy of the tribals during the pre-colonial period would not have made any serious impact on forest resources. An approximate equilibrium in the natural resource base existed between the years 800 and 1800.[26]

Mixed cultivation of two to five crops was followed.[27] Labour requirements were met mainly by family members; indeed, there was no labour market. Payments were not made but everyone partook in the harvest feast.

Individual ownership of land did not exist, and the community managed the land collectively: the forests were a common property resource.[28] Until the intervention of external forces, most of the hills were isolated from mainstream society, and the tribals had only minimal interaction with others in order to obtain essential commodities.[29] The present economy of the tribals is completely different from its previous incarnation, before the intervention of external forces in their traditional socio-economic and cultural system.

Given the nature of the tribal economy and livelihood, one can ask whether there was any deforestation, and if so to what extent it occurred and how sustainability maintained. Questions like this are important if we are to gain a better understanding of the earlier tribal economic system and resource management.

In the forest areas, the tribals' livelihood was also at the irreducible minimum subsistence, mainly managed with forest resources. At the same time, sustainability was maintained owing to the following reasons: tribals' needs were very limited, the density of population was low and forest resources were not commercialised. In short, the low population density allowed the cultivation and collection of forest resources without commercial motives, which ensured the sustainability of environment and ecosystem until the close of the eighteenth century in different hill regions of south India.

Given this scenario, one can also ask whether non-tribals were responsible for deforestation, and if so to what extent. If not, how did they manage forest resources? Even on the plains, villagers managed their livelihood among themselves according to the natural resources that were available. They did not depend on outsiders for their livelihood until the late eighteenth century.[30] On the plains, every village invariably had common property resources, consisting of pastoral and uncultivable land, which was used for the collection of fuel-wood and the grazing of cattle. In addition, only a few industries were located on the plains, which used the fuel-wood, by and large from the plains themselves. For example, iron-making, sugar-making and brick-making industries existed in different parts of south India at the close of the eighteenth century. The iron-making industries supplied agricultural implements, tools for carpenters and smiths, and iron boilers for sugar-making, as well as domestic utensils.[31] These industries were managed with the locally available fuel-wood, which was burnt for charcoal to be used by an individual village. Forest resources were not used for commercial purposes either by the tribals or by the plains people until the close of the eighteenth century.

To conclude, until the intervention of external/market forces, the tribals and other forest users used the forest resources mainly for their livelihood at the minimum irreducible level, which ensured sustainability. In other words, the tribals did not extract forest resources for commercial purposes. Likewise on the plains, the non-tribal people and even the limited number of traditional industries did not cause deforestation in any part of south India. This may be adduced from the extensive forest resources in different parts of south India that existed at the close of the eighteenth century. In a nutshell, both the tribals and others used the forest resources only for their livelihood, which ensured the sustainability of environment and ecology throughout south India until the close of the eighteenth century.

Colonial Policy in Respect of the Forests (1792–1882)

From the late eighteenth century, external/market forces appropriated the tribal managed/controlled forest resources for different commercial purposes, including the extraction of timber for the shipping fleet, the extraction of sandalwood, the establishment of tea and coffee plantations, the extraction of forest resources for iron-making industries and the establishment of railways in different parts of south India. Restrictions were imposed upon the tribals to limit their use of the forest resources. Let us see how colonial forest policy appropriated forest resources during the pre-Forest Act period.

Timber Extraction for the Fleet

The British were forced to find timber in India because of a scarcity of timber for the construction of the fleet in England that occurred during the latter part of the eighteenth century.[32] Early colonial rule mainly focused on extracting teak to meet the navy's demands.[33] Subsequently, the East India Company proclaimed that teak trees belonged to the Company and prohibited their unauthorised felling.[34] As pressure from Britain to ensure future supplies to the navy increased, the Company appointed a Conservator of Forests in 1807 'to preserve and improve the protection of teak and other timber suitable for shipbuilding'.[35] Soon a timber monopoly was established throughout the Malabar and Travancore regions, and 'destruction of the more accessible forests increased during the early days of British occupation'.[36] These measures were contested both by proprietors and merchants, which led to the abolition of the conservator post.[37] In short, early colonial rule extracted forest resources mainly to meet the demand from their own country. Any attempt to protect the forest was not really undertaken to conserve it but to ensure future supply, particularly to meet the British demands.

A large quantity of timber and wood was exported from south India to Arabia, Ceylon and other countries, besides a large quantity via the inter-provincial trade to Bombay, Scinde, Bengal and Burma.[38] For instance, about Rs.22,985,179 worth of timber and wood was exported between 1855/1856 and 1875/1876 (Table 2.1). This shows that forest resources were used by other countries for commercial purposes during the pre-Forest Act colonial regime.

Extraction of Sandalwood

From the early nineteenth century, a large quantity of sandalwood from the hills/forest was extracted to export to other countries for commercial motives.[39] In fact, the colonial government encouraged middlemen to export sandalwood by providing the required facilities. For example, about 62,984–11¾ *mounds* of sandalwood were delivered from Coimbatore district alone to the Board of Trade between 1799 and 1809,[40] this being worth 104,915 *Star Pagodas*,[41] 9 *Fanams*,[42] 8 *Cash*.[43]

Table 2.1 Value of timber and wood exports from the Madras Presidency, 1855/1856–1875/1876

Year	*Value (in rupees)*	*Year*	*Value (in rupees)*
1855–1856	736,117	1866–1867	930,878
1856–1857	853,704	1867–1868	979,671
1857–1858	882,080	1868–1869	1,307,459
1858–1859	884,444	1869–1870	1,228,689
1859–1860	847,820	1870–1871	1,013,878
1860–1861	931,317	1871–1872	1,022,367
1861–1862	1,237,475	1872–1873	1,014,176
1862–1863	1,144,023	1873–1874	1,028,803
1863–1864	1,570,559	1874–1875	1,065,906
1864–1865	1,606,538	1875–1876	1,045,109
1865–1866	1,654,166		

Source: Maclean, *Manual of the Administration of the Madras Presidency*, Government Press, Madras, 1885, p. 344

After the colonial intervention, these sandalwood trees were extracted for commercial purposes and even exported to other countries, such as China. For instance, in Salem district alone, revenue from sandalwood was about 3,776 *Star Pagodas*, 30 *Fanam* and 40 *Cash* in 1800/1801.[44] The Collector reported that a significant amount of sandalwood, about 200–400 *candies*,[45] could be supplied from the district annually.[46] Later, the right to cut sandalwood was leased out to private contractors. In about 1836, the government attempted to ascertain the amount of sandalwood in different regions, and discovered that the precious trees were plentiful throughout Salem district.[47] Subsequently, through the lease/rent of sandalwood cutting rights in different hills, the colonial government earned a large amount of revenue: for instance, in the decade between 1862/1863 and 1875/1876, about Rs.147,537 was realised (Table 2.2). In other words, about a quarter of the Forest Department's total receipts came from sandalwood in the Salem and Baramahal region.

Coffee and Tea Plantations

Coffee and tea plantations were established in various hills of south India during the early nineteenth century, particularly between the 1830s and 1850s. Coffee plantations were largely found in the highlands of the different districts of the Madras Presidency, Vizagapatanam, Madurai, Tinnevelly, Coimbatore, Nilgiris, Salem, Malabar and South Canara, and the Princely States, Cochin and Tiruvancore.[48] Coffee cultivation began in the 1830s. As of 1883, 17,872 coffee plantations had been established,

Table 2.2 Revenue from sandalwood and other sources from Salem and Baramahal regions, 1862/1863–1875/1876 (in Rupees)

Year	*Revenue from sandalwood*	*Year*	*Revenue from sandalwood*
1862–1863	8573	1869–1870	14,550
1863–1864	9758	1870–1871	9153
1864–1865	13,068	1871–1872	4594
1865–1866	10,292	1872–1873	4441
1866–1867	15,261	1873–1874	7502
1867–1868	17,110	1874–1875	4754
1868–1869	15,929	1875–1876	12,552

Source: Letter from Colonel R.H. Beddome, Conservator of Forests to J.H. Garstin, Acting Secretary to Government, Revenue Department, dated 22 December 1876 (Chennai: Tamil Nadu State Archives)

Table 2.3 Coffee cultivated area in the Madras Presidency and Princely States, 1883

Name of the district	*No. of coffee plantations*	*Cultivated area (acres)*
Madras Presidency		
Vizagapatanam	9	6.5
Madurai	3355	5378
Tinneveli	51	2719
Coimbatore	16	1321
Nigiris	459	35,128
Salem	331	10,769
Malabar	13,568	52,965
South Canara	10	72
Princely States		
Cochin	17	8251
Travancore	56	11,386
Total	17,872	127,996

Source: Maclean, *Manual of the Administration of the Madras Presidency*, Government Press, Madras, 1885, pp. 403–404

covering an area of 127,996 acres both in different parts of the Madras Presidency and the Princely States (Table 2.3). The colonial government encouraged coffee and tea cultivation by providing various subsidies and concessions.[49]

A large number of coffee products were exported to other countries, and the export quantity progressively increased in every year during the nineteenth century. For instance, in 1855/1856, about 96 lakh lbs, worth about Rs.9 lakh was exported, and this increased respectively to 367 lakh lbs and Rs.136 lakh in 1875/1876. On the whole, about 5538 lakh lbs of coffee worth Rs.1352 lakhs were exported during the two decades 1855/1856 to 1875/1876 (Table 2.4). In other words, the growth rate of export increased by 326% for the quantity and 1426% for the value between the periods 1855/1856 and 1875/1876. Coffee was exported predominantly to the United Kingdom and France, besides Austria, Turkey, Egypt, the United States of America, Arabia, Ceylon, Persia, Victoria and foreign ports in India as well as other ports of the Indian subcontinent.[50]

Tea estates were found in the Vizagapatanam, Madurai, Nilgiris and Malabar districts of the Madras Presidency and Travancore in the Princely States. In 1883, about eighty-six estates with 11,874.75 acres of tea and twenty-six estates with 1005 acres were found in Travancore (Table 2.5).

Table 2.4 Trends of coffee export in the Madras Presidency, 1855/1856–1875/1876

Year	*Quantity (lbs)*	*Value (in rupees)*	*Year*	*Quantity (lbs)*	*Value (in rupees)*
1855–1856	8,601,464	892,167	1866–1867	34,527,695	7,813,813
1856–1857	8,869,609	944,446	1867–1868	17,374,766	4,191,785
1857–1858	7,714,355	867,044	1868–1869	35,659,638	8,058,373
1858–1859	11,372,338	1,230,803	1869–1870	47,411,508	10,765,221
1859–1860	14,578,228	1,875,054	1870–1871	37,353,359	8,821,118
1860–1861	18,571,414	3,241,699	1871–1872	35,484,209	8,284,862
1861–1862	20,960,198	4,710,369	1872–1873	56,889,888	13,834,164
1862–1863	20,269,004	5,355,581	1873–1874	41,548,950	11,276,542
1863–1864	27,333,127	6,555,671	1874–1875	41,179,712	15,191,418
1864–1865	31,424,319	7,684,939	1875–1876	36,652,965	13,613,617

Source: Maclean, *Manual of the Administration of the Madras Presidency*, Government Press, Madras, 1885, p. 336

Table 2.5 Tea cultivated area in the Madras Presidency and Princely States, 1883

Name of the district	*No. of tea plantations*	*Cultivated area (in acres)*
Madras Presidency		
Vizagapatanam	1	42
Madurai	2	3
Nigiris	78	11,764
Malabar	5	66
Princely States		
Travancore	26	1006
Total	112	12,881

Source: Maclean, *Manual of the Administration of the Madras Presidency*, Government Press, Madras, 1885, p. 404

These plantations were largely concentrated in the hill areas, where the tribal settlements existed. For example, in Tamil Nadu, coffee and tea plantations were established in the Shervaroys and the Nilgiris during the early nineteenth century. The planters encroached/alienated a large quantity of tribal occupied land and the common property resources. In addition, they disturbed the traditional tribal system.[51] In 1885, approximately 108,358 acres of land were under coffee plantations in the erstwhile Madras Presidency.[52] Coffee and tea plantations were encouraged by the administration through the provision of concessions. These included the exemption of land rent for the first five years and a meagre land rent of 1 rupee per acre per annum for the subsequent sixteen years of the lease period. In most cases, tribals were harassed and badly disturbed by the planters who wanted to grab their land.[53] Consequently, a sizeable number of tribals became landless and plantation labourers. In addition, the government imposed several restrictions on tribals' rights over their lands and common property resources.[54] Hence, their earlier modes of production, which were directly related to natural resources, were disturbed from the early nineteenth century throughout south India.

Tea was exported to other countries from early in the second half of the nineteenth century. Initially, a large quantity was exported: for instance, in 1855/1856 about 43,426 lbs with a value of Rs.34,376. However, after a year the export quantity dwindled, and in the decade between 1858/1859 and 1868/1869 tea exports were completely neglected. Exports increased progressively in the subsequent period, though: between 1869/1870 and 1882/1883 about 2,035,906 lbs of tea worth Rs.2,269,304 were exported to other countries (Table 2.6). This clearly shows that the colonial administration encouraged the establishment of plantations in different hill regions, mainly for export to the other countries, and consequently a substantial amount of forest was destroyed during the early nineteenth century in south India.

Table 2.6 Trends of tea export in the Madras Presidency, 1855/1856–1875/1876

Year	*Quantity (lbs)*	*Value (in rupees)*	*Year*	*Quantity (lbs)*	*Value (in rupees)*
1855–1856	43,426	34,376	1874–1875	93,918	114,383
1856–1857	79,915	57,461	1875–1876	120,924	146,272
1857–1858	13,699	13,699	1876–1877	147,132	166,988
1858–1859	53	70	1877–1878	183,178	193,083
1869–1870	6166	9162	1878–1879	204,630	211,753
1870–1871	17,883	18,065	1879–1880	208,247	233,904
1871–1872	33,579	45,594	1880–1881	266,942	286,722
1872–1873	43,426	52,417	1881–1882	319,752	372,738
1873–1874	80,581	89,166	1882–1883	309,548	329,057

Source: Maclean, *Manual of the Administration of the Madras Presidency*, Government Press, Madras, 1885, p. 346

Iron-Making and Sugar-Boiling Industries

Iron deposits are extensively found in south India. 'In Salem, the Northern Circars, Ceded Districts, and Mysore, iron is found largely in the ore, while in Coimbatore and Malabar the black sand brought down by streams is rich in the metal.'[55] Until the colonial intervention, traditional iron-making industries supplied agricultural implements, carpenters' and smiths' tools, iron boilers for sugar-making, domestic utensils and many other requirements of the people.[56] These industries usually employed from three to ten workers, and in some cases up to twenty.[57] Traditional iron-making industries, though spread over different parts of the country, mostly depended on locally available fuel-wood to make charcoal, mainly produced in individual villages.

After colonial intervention, the colonial government encouraged British-owned iron-making industries and provided some exclusive privileges like mining rights and tax exemption on fuel-wood. From the early nineteenth century, the colonial government emphasised the extraction of iron ore. In 1824, Thomas Munro observed that 'the richness of the ores of Salem and other districts have been known above thirty years, without anything having been done to extend their produce, and he suggested that the State has to extend the support for extracting the resources'.[58] In 1824, Mr Heath, who was working in the Madras Civil Service, asked the government for the exclusive right to iron-making in India and a lease for the right to cut fuel in the government wasteland.[59] In fact, Munro and other Company members recommended Mr Heath's proposal, with the support of necessary facilities to establish the iron-making industries. But the Court of Directors declined to accept the proposal.[60] After some years, in 1830, the Court of Directors granted Mr Heath the exclusive privilege to establish iron-making industries in the Madras Presidency.[61] He established the iron-making industries at Porto Nova in South Arcot district with the help of the government.

In 1833, Mr Heath applied to the government for further financial assistance as well as other support. Considering the cost of production when compared with other countries, the government extended its support to the iron-making industry.[62] 'Accordingly in 1833 Mr. Heath was granted exclusive mining rights in Salem, Coimbatore, South Arcot, Malabar and Canara and gave permission to cut fuel in Tanjore, Tiruchinopoly and South Arcot with exemption from various duties for 21 years.'[63] In addition to this, the duty on Indian iron in England was reduced and all the materials required for Porto Nova were allowed to pass duty free.[64] In 1853, the colonial government granted permission to the East India Iron Company for the exclusive right of extracting ore from government wastelands in South Arcot, Salem, Malabar, Canara and Coimbatore for 30 years on the payment £500 rent per annum.[65] In addition to the native iron-smelting industries and the British iron-making company, the sugar-boiling industries in the different districts of the Madras Presidency consumed a large quantity of fuel-wood during the early colonial period.[66] This clearly indicates that the colonial government encouraged the iron-making industries by extending all kinds of concessions, leading to the decline of forest resources in south India during the early nineteenth century.

Railways

Since the introduction of railways in the Madras Presidency in 1853, the length of railway track steadily increased. The first line, between Veyasarpandy and Walajah Road (Arcot), 63 miles in length, was opened on 1 July 1856 by the Madras Railway Company. The track length had increased to 690 miles by 1866 and to 1436 miles by 1877 (Table 2.7). In other words, the growth rate was over 2179% within two decades.

Table 2.7 Expansion of railway line in the Madras Presidency, 1856–1877

Year	*Total route (miles)*	*Year*	*Total route (miles)*
1856	63	1867	772
1857	65	1868	825
1858	81	1869	863
1859	96	1870	893
1860	137	1871	1018
1861	236	1872	1018
1862	312	1873	1022
1863	522	1874	1044
1864	527	1875	1172
1865	650	1876	1384
1866	690	1877	1436

Source: *Statistical abstract relating to British India* (various years)

Table 2.8 Quantity of wood purchased for fuel by Madras Railways, 1873–1877

Years	*Total quantity purchased (in tons)*	*Total wood consumed exclusive of small wood for lighting (lbs) (tons in bracket)*
1873	44,693	118,841,314 (52,936.00)
1874	55,107	117,959,691 (52,543.29)
1875	57,466	139,582,000 (62,174.61)
1876	77,731	162,029,750 (72,173.61
1877	119,924	263,091,895 (117,190.15)
Total	354,921	801,504,650 (357,017.66)

Source: *G.O.No.373 PWD (Railway), 26 April 1878*, Tamil Nadu state Archives, Chennai

A large quantity of forest resources were extracted owing to the establishment of railways.[67] For example, in 1878 the Madras Railways agent stated that the approximate requirement of wood for fuel was about 91,000 tons.[68] Owing to the laying of railways, a large number of valuable trees were cut down throughout south India. For example, in 1859/1860 about 245,743 berths were supplied from Salem and Baramahal region alone.[69] In addition to the berths, a large quantity of wood was purchased for fuel by Madras Railways. About 354,921 tons of wood were purchased and about 801,504,650 lbs or 357,017.66 tons were consumed, exclusive of small wood for lighting during the period 1873–1877 (Table 2.8). Further, forest conservation was initiated in different districts of the Madras Presidency to meet future fuel-supply needs of Madras Railways. In short, large quantities of forest resources were extracted for the railways in different parts of the Madras Presidency during the second quarter of the nineteenth century.

Reserve Forest Prior to the Forest Act

A large extent of forest area was brought into the reserve forest from 1867, prior to the Madras Forest Act 1882, not because of conservation but to ensure future commercial requirements.[70] Between 1867 and 1881, about 756,562 acres were brought into the reserve forest by the colonial administration, either through Government Order or by the Board of Revenue (Table 2.9).

Table 2.9 Area under the reserve in the Madras Presidency prior to the Madras Forest Act, 1867–1881

District	*Year of notification*	*Estimated areas in acres*
Godavari	1880	43,520
Cuddappah	1870–1881	113,286
North Arcot	1871–1879	14,381
South Arcot	1869–1880	95,757
Tiruchinopoly	1871	4166
Madurai	1878–1881	150,675
Tinnevelly	1881	183,558
Coimbatore	1879	92,941
Nilgiris	1878–1880	18,074
Salem	1867–1880	11,398
Malabar	1872–1874	28,806

Source: Brandis, *Suggestions Regarding Forest Administration in the Madras Presidency*, Government Press, Madras, 1883, p. 16

Revenue from the Forest

From the early nineteenth century, the Forest Department extracted a sizeable amount of revenue from the forests, either through selling forest resources directly or by leasing out the rights of collection of forest produce in different parts of the Madras Presidency. Although there is no evidence of a separate account for this during the early nineteenth century, the available data clearly indicates that the revenue generated from the forest was not negligible. In 1868, about £42,966 was collected from forest resources, and between 1868 and 1877, about £415,258 was collected from the forest in the Madras Presidency (Table 2.10). This may be attributed to commercialisation of forest resources at a time when conservation was gaining emphasis in the Madras Presidency. In other words, if the tribals made use of the forest resources for their livelihood it was caricatured as destruction, but if the state acted in a similar fashion it was portrayed as protection and development. From the early nineteenth century, different forest resources were extracted for commercial purposes. Different incentives were provided for this at different times. Commercial exploitation of forest resources certainly affected the environment and the tribals, which we will see in the next section.

Table 2.10 Amount of forest revenue collected from the Madras Presidency, 1868–1877

Year	*Forest revenue (in £)*	*Forest revenue (in rupees)*[a]
1868	42,966	429,660
1869	39,133	391,330
1870	49,855	498,550
1871	33,979	339,790
1872	41,886	418,860
1873	41,854	418,540
1874	44,983	449,830
1875	36,354	363,540
1876	42,772	427,720
1877	41,476	414,760

Source: Statistical Abstract Relating to British India from 1867/1868 to 1876/1877

[a]One £ = 10 rupees

Consequences of the Forest Policy on the Environment and Forest-Dwellers

As explained earlier, forest resources in the hill country were never exploited for commercial purposes by the tribal people, who made judicious use of them and ensured their sustainability until the colonial intervention took place. '[During the pre-British period] scholars generally agree that the scale of the logging was too small to cause any widespread ecological damage.'[71] In this section, we will examine the commercial exploitation of forest resources during the pre-Forest Act period and the impact this had on the environment and livelihood of the tribals and other people.

Decline of the Environment

The earliest travel account, by Buchanan (1807), does not give a detailed account of the different hills of the Madras Presidency. Yet it was observed that the hills had an abundance of trees, particularly on their slopes.[72] Other sources, such as folklore and oral histories, are limited and include little information on the changes in landscape in these hill regions. However, the archival sources provide sufficient evidence for a decline in the environment. Colonial intervention prior to the introduction of the Forest Act witnessed a large-scale export of sandalwood by the government to other countries, the destruction of forests to establish coffee/tea plantations and the plunder of timber and other forest resources for the industries and railways. It is clear that commercialisation, which commenced in the late eighteenth century, gained ascendancy with the government adopting different strategies before the introduction of the Madras Forest Act in 1882.[73]

From the early nineteenth century, the establishment of plantations in different hills led to the destruction of forests, which in turn impacted negatively on the environment. In fact, the colonial administration actively encouraged and promoted British planters by providing concessions.[74] In due course, the planters encroached on vast tracts of common property resources.

The traditional iron-smelting and sugar-boiling industries that depended on forests for fuel-wood were limited. In contrast, the British iron-making industries favoured by the administration extensively availed themselves of their exclusive privileges, with the British sugar-boiling industries also continuing to consume very large quantities of fuel-wood—speeding deforestation to some extent in the early nineteenth century.[75] Indeed, these industries between them extracted a large quantity of forest resources in different regions of south India.

The establishment of Madras Railways in the latter part of the nineteenth century provided an additional momentum to the extraction of forest wealth, especially that of timber and fuel-wood.[76] The revenue expenditure was much lower than receipts.[77] This clearly indicates that the forest resources were extracted for the commercial purpose, which implies a decline in environmental conditions in the Madras Presidency.

Since the early nineteenth century the colonial government has extracted forest resources for different commercial purposes and also cut them down in order to establish plantations. Resources satisfied Company requirements but were also exported for commercial gain. In fact, the colonial government has encouraged the exploitation of forest resources by providing concessions and privileges since the early nineteenth century. Indeed, the extraction of forest resources and ensuring that their supply met present and future requirements was the main agenda of the colonial government during the pre-Forest Act regime.

Decline of Tribals

From the early nineteenth century the traditional tribal system in south India disintegrated and the tribals' economic conditions deteriorated. The tribal traditional administrative set-up was replaced by the colonial government. The tribals were also displaced from their land, which led to a decline in the tribal system.

Disintegration of Tribal Management

Since the colonial intervention took place, revenue collection had been settled with the tribal headmen. The tribal headman used to collect revenue from the tribals and pay the government. But between 1829 and 1854 the role of tribal headman was withdrawn by the colonial administration, and the hills were brought under direct management on one pretext or another.[78] The government brought under *amani* management part of the Attur-Kolli hills in 1828/1829, Aranuthumalai in 1836/1837, the Namakkal-Kolli hills in 1840/1841, the Shervaroy hills in 1842/1843 and the remaining portion of the Attur-Kolli hills in 1853/1854.[79] The traditional rights of the headmen and the *guru* (priest) were usurped in the process.[80] Traditional tribal administration was withdrawn so that the commercialisation of forest resources could be executed without any hindrance.

Non-Tribal Settlement

In the early nineteenth century, before colonial rule, the non-tribal presence in the hill areas was insignificant because the plain people dreaded travelling to the hills for fear of catching contagious diseases. After the colonial intervention, the British dared to settle in different hill areas in south India, and they were followed by native non-tribals. Until the 1871 Census, there was no separate data for non-tribals in the hills, but we can observe the trends regarding non-tribal settlement in different hills of south India. Data pertaining to the different hills of Salem and Baramahal regions shows that non-tribal settlement increased during the early nineteenth century.[81] According to the 1871 Census, about 4026 persons or about 10.21% of the total population in the five major hills of Salem and Baramahal regions were found to be non-tribals. Of them, 2649 persons or about 65.11% were *Pariahs* (a Scheduled Caste under the Indian constitution), who were bonded labourers in the plains (Table 2.11).[82] Only

Table 2.11 Different castes of non-tribals in different hills of Salem and Baramahal regions of the Madras Presidency, 1871

Hills	*Total population*	*Non-tribals*	%	*Pariahs*	%	*Chetties*	%	*Muslims*	%	*Others*	%
Kalrayan hills	6626	298	4.50	18	6.04	–	–	3	1.01	276	92.62
Pachamalai	2166	13	0.60	–	–	–	–	–	–	13	100
Attur-Kolli hills	5533	574	10.37	53	9.23	–	–	1	0.17	520	90.59
Namakkal-Kolli hills	9296	852	9.16	635	94.53	1	0.12	–	–	216	25.35
Shervaroy hills	10,745	2289	21.30	1943	84.88	6	0.26	22	.96	318	13.89

Sources: *Census Statement of Population of 1871, Salem District*, pp. 142–143, 148–151, 160–161 and 290–291

seven persons or about 0.17% were *Chetties* (a business community), 26 persons or about 0.64% were Muslims and the remaining 1343 persons or about 34% belonged to the so-called Other Backward Class (OBCs). In the Shervaroy hills, about 2289 persons or about 21.30% of the total population in the hills were found to be non-tribals. Coffee plantations had been established there in 1825, and around ten Europeans were engaged in the venture by the end of the first half of the nineteenth century.[83]

According to Shortt (1864), twenty-seven planters were engaged in coffee cultivation, of whom twenty-five were British and the rest were non-tribals. Interestingly, about 100 persons including children belonged to European stock among the settlers in the Shervaroy hills from the 1870s.[84] After the colonial intervention, the non-tribal settlement in the hill areas increased in different regions of south India.

Tribal Land Alienation

Non-tribal settlement led to tribal land alienation in different hills of south India during the early nineteenth century. It is very obvious where plantations were established from the second quarter of the nineteenth century. For instance, in the Shervaroy hills alone about 9210 acres were under coffee cultivation in the 1870s. Invariably these lands were taken over from the tribals. Interestingly, about 7776 acres or 85% of the area under plantations was held by the British while non-tribal natives possessed the remaining.[85] In the other hills as well the non-tribals alienated the tribal occupied land. The nature and extent of tribal land alienation in different hills can be seen by the deterioration in occupational status.[86]

Decline of Economic Status

Non-tribal settlement and alienation of tribal land led to a decline in the economic status of the tribals throughout south India. Until the advent of the colonial administration and even afterwards for some time, the tribal economy was associated with traditional methods of cultivation and collection of forest produce.[87] In fact, the tribals were classified under the occupational category cultivators in 1836.[88] Owing to the intervention of the British settlers and other non-tribals, and with the introduction of coffee plantations and reserve forests, their earlier occupational status deteriorated from the early nineteenth century. For instance, about 12,021 (35%) of the people in the five major hills were classified as workers, of

Table 2.12 Occupational status of the tribals in Salem and Baramahal regions, 1871

Name of the hills	*Total population*	*Total workforce*	*% of col. 3 to 2*	*Labourers*	*% of labourers to total workforce*
Kalrayan hills	6626	2321	35.03	619	26.67
Pachamalai	2166	795	36.70	27	3.40
Attur-Kolli hills	5533	2231	40.32	71	3.18
Namakkal-Kolli hills	9296	3658	39.35	929	25.40
Shervaroy hills	10,745	3016	28.07	857	28.42
Total	34,366	12,021	34.98	2503	20.82

Sources: *Census Statement of Population of 1871, Salem District*, pp. 142–143, 148–151, 160–161 and 290–291

which 2503 (21%) were classified as agricultural labourers (Table 2.12). More and more tribals became either agricultural labourers or dependent upon agricultural and allied activities. It is evident that in the Shervaroy hills, where the greatest extent of tribal lands was alienated by the planters, the result was a large proportion of agricultural and plantation labourers.[89] Further, where the proportion of non-tribals settlement was high the proportion of labourers was also high, showing the decline in their occupational status. In other words, the evidence clearly suggests that the tribal managed forest resources were appropriated for different commercial purposes and a large number of non-tribals settled in the hill areas, which eventually led to a decline in their livelihood.

Conclusion

Until the colonial period (1792), the sustainability of forest resources was not affected by the tribal people or by the other forest users, as exploitation did not take place for commercial purposes. Until the intervention of external/market forces, the tribal economy remained a subsistence economy, managed by custom and limited to the collection of forest produce and traditional cultivation. Being a non-market economy was its most distinctive feature. The community members, without any intervention from external forces, managed the resources among themselves. An inevitable consequence of external intervention was a drastic alteration in the non-market tribal economy, besides weakening the common property resources particularly during the early nineteenth century. In fact, the penetration of market forces was not confined to the economic sphere alone; permeated the entire socio-economic, cultural and administrative set-up of the tribal community.

From the early nineteenth century, forest resources were exploited for commercial purposes by the colonial government. No constructive measures were initiated to conserve these resources until it was felt they were becoming exhausted. In other words, the conservation of forest resources was initiated on a small scale that aimed to meet present as well as future requirements. One can see the same trend even in the post-Forest Act period, during which forests were demarcated, surveys were undertaken, working plans were prepared and fire protection measures were initiated. But we should ask why these measures were initiated. It 'Even the initiatives directed towards environmental protection were indeed guided by the need to sustain the resources for future exploitation. In order to ensure sustainability, the Forest Act had imposed a whole range of restrictions on the tribals and other forest users since the late nineteenth century.'[90] Further, this was recently recognised by the government. While introducing the tribal bill The Scheduled Tribes (Recognition of Forest Rights) Bill, 2005, it was accepted 'to recognise and vest the forest rights and occupation in forest land in forest dwelling Scheduled Tribes who have been residing in such forests for generations but whose rights could not be recorded; to provide for a framework for recording the forest rights so vested and the nature of evidence required for such recognition and vesting in respect of forest land'.[91] Further, 'the forest rights on ancestral lands and their habitat were not adequately recognized in the consolidation of State forests during the colonial period as well as in independent India resulting in historical injustice to the forest dwelling Scheduled Tribes who are integral to the very survival and sustainability of the forest ecosystems'.[92] It was also substantiated that even during the joint forest management regime, when a large number of tribals and other forest users were involved in forest management during the last two decades (1980–2000) in Andhra Pradesh. forest-related offences were executed by the timber mafia.[93] Hence, one can argue that the conservation measures of the colonial government in south India during the early nineteenth century were not really to preserve the forest resources but to meet commercial interests. It is clear that establishing coffee plantations during the nineteenth century damaged the ecology, resulting in environmental decline and a loss of livelihood for the tribal communities in the Shervaroy hills of the Madras Presidency.[94]

Further, it may be argued that the focus of colonial policy was on extraction of forest resources in utter disregard of the impact on the environment and the resultant hardships caused to the tribal populace.[95] The colonial government changed its policy not with the aim of improving the living standards of the tribals or conserving forest resources but in order to extract them by controlling certain geographical areas. Besides encouraging the commercialisation of forests and non-tribal settlements, it led to the disintegration of the traditional tribal system as well as the tribal forest-oriented economy. To cut a long story short, the British colonial forest policy did not accord due importance to the environment, since it was concerned merely with the commercialisation of forest resources during the pre-Forest Act colonial era.

Notes

1. Velayutham Saravanan, 'Colonial Commercial Forest Policy and the Tribal Private Forests in Madras Presidency: 1792–1881', *Indian Economic and Social History Review*, 40(4), 2003, pp. 403–423; Velayutham Saravanan, 'Colonialism and Coffee Plantations: Decline of Environment and Tribals in Madras Presidency during the 19th Century', *Indian Economic and Social History Review*, 41(4), 2004, pp. 465–488.
2. Ramachandra Guha, 'Forestry in British and Post–British India: A Historical Analysis' *Economic and Political Weekly*, 18(45–46), 1983, pp. 1886 and 1892; Ramachandra Guha and Madhav Gadgil, 'State Forestry and Social Conflict in British India' *Past and Present* 122, 1989, pp. 148–157; Marlene Buchy, 'British Colonial Forest Policy in South India: An Unscientific or Unadopted Policy?' in Richard H. Grove, Satpal Sangwan and Vinita Damodaran (eds) *Nature and the Orient: The Environmental History of South and Southeast Asia*, Delhi: Oxford University Press, 1998, p. 669; Velayutham Saravanan, 'Commercialisation of Forest, Environmental Negligence and Alienation of Tribal Rights in Madras Presidency: 1792–1882,' *Indian Economic and Social History Review*, 35(2), 1998, pp. 125–146; Saravanan, 'Colonialism and Coffee Plantations'; Dhirendra Datt Gangwal, 'Commercialisation of Forests, Timber Extraction and Deforestation in Uttaranchal, 1815–1947', *Conservation and Society*, 3(1), 2005, pp. 110–133; David Arnold and Ramachandra Guha, 'Introduction: Themes and Issues in Environmental History of South Asia' in David Arnold and Ramachandra Guha (eds), *Nature, Culture, Imperialism: Essays on the Environmental History of South Asia*, Delhi: Oxford University Press, 1995, p. 11; Neeladri Bhattacharya, 'Pastoralists in a Colonial World' in David Arnold and Ramachandra Guha

(eds), *Nature, Culture, Imperialism: Essays on the Environmental History of South Asia*, Delhi: Oxford University Press, 1995, pp. 49–85; Ajay Skaria, 'Timber Conservancy, Desiccationism and Scientific Forestry: The Dangs 1840s–1920s' in Richard H Grove, Satpal Sangwan and Vinita Damodaran (eds), *Nature and the Orient*, Delhi: Oxford University Press, 1998; Ajay Pratap, *The Hoe and the Axe: An Ethnohistory of Shifting Cultivation in Eastern India*, Delhi: Oxford University Press, 2000; Mahesh Rangarajan, *Fencing the Forest: conservation and ecological change in India's central province 1860–1914*, Delhi: Oxford University Press, 1996; Mahesh Rangarajan, 'Protection, Desiccation and Forest Management in the Central Provinces 1850–1930' in Richard H Grove, Satpal Sangwan and Vinita Damodaran (eds), *Nature and the Orient*, Delhi: Oxford University Press, 1998, p. 589; David Baker, 'State polity, the market economy, and Tribal decline: the central provinces, 1861–1920', *Indian Economic and Social History Review*, 28(4), 1991, pp. 341–370; N.C. Saxena, *Forest Policy in India*, New Delhi: World Wide Fund for Nature–India, and London: International Institute for Environment and Development, 1999.

3. Richard H Grove, Green Imperialism: Colonial expansion, tropical island Edens and origins of environmentalism 1600–1860, Delhi: Oxford University Press, 1995.
4. Ibid.
5. Saravanan, 'Commercialisation of Forest, Environmental Negligence and Alienation of Tribal Right'; Saravanan, 'Colonial Commercial Forest Policy and the Tribal Private Forests'; Saravanan, 'Colonialism and Coffee Plantations'; Gangwal, 'Commercialisation of Forests, Timber Extraction and Deforestation'; Saxena, 'Forest Policy in India'.
6. Akileshwar Pathak, *Law, Strategies, Ideologies: Legislating Forests in Colonial India*, Delhi: Oxford University Press, 2002.
7. B.K. Mishra, B.A. Bhanu, K.N. Reddy, M Sethuraman and T.S. Natarajan, *Tribes of Southern Region [A Select Bibliography]*, New Delhi: Inter–India Publications, 1968; B. Janarthan Rao, *Land Alienation in Tribal Areas.* Kakatiya School of Public Administration Publication, Warangal, 1987.
8. S. Ramanathan, *Tribal Welfare in Salem District: The Role of Government and Voluntary Organisations*, Madras: Sangam Publishers, 1977; A. Jothimurugan, *Tribal Development in Tamil Nadu*, unpublished M.Phil dissertation, Madras: University of Madras, 1985; A. Jothimurugan, *Socio–Economic Impact of Tribal Development Schemes in Tamil Nadu: A Comparative Studies of Kalrayan Hills of Salem and South Arcot Districts*, unpublished Ph.D dissertation, Madras: University of Madras,1989; Velayutham Saravanan, *Working of Nutritious Noon Meal Scheme in Kalrayan Hills: With Special Reference to Peddanaicken Palayam Block in Salem District*, unpublished M.A., dissertation, Tiruchirappalli: Bharathidasan University, 1987; Velayutham Saravanan, *The Socio–*

Economic Conditions of Scheduled Tribes: With Special Reference to Malaiyalis of Peddanaicken Palayam Block in Salem District, unpublished M.Phil., dissertation, Madras: University of Madras,1989.

9. Saravanan, 'Commercialisation of Forest, Environmental Negligence and Alienation of Tribal Right'; Saravanan, 'Colonial Commercial Forest Policy and the Tribal Private Forests'; Saravanan, 'Colonialism and Coffee Plantations'; Saravanan, 'Tribal Revolts in India with reference to Salem and Baramahal Districts of Madras Presidency' pp. 67–81; Saravanan, 'Commercial Crops, Alienation of Common Property Resources and Change in Tribal; Velayutham Saravanan, 'Thamizhaga Malaival Makkalin Nila Urimai', *Uzhavan Urimai*, (Land Rights of Tribals in Tamil Nadu) [Special Issue] 1999, pp. 53–57 (in Tamil); Saravanan, 'Tribal Land Alienation in Madras Presidency during the Colonial period: 1792–1947,' pp. 73–104.
10. C.T. Kurien, *The Economy: An Interpretative Introduction*, New Delhi: Sage Publications, 1992, p. 30.
11. Ibid., p. 31.
12. Approximately covering the period between three centuries before and after the Christian era.
13. Busnagi Rajannan, *Salem Cyclopedia*, Salem: Institute of Kongu Studies, 1992, pp. 131–32.
14. P. Radjn, *The World of Primitive Man*, Henry Schuman, New York, 1953; Stephen Viljoen, *The Economic of Primitive People.* Staples Press Ltd., New York, 1936, p. 11; Kurien, *The Economy: An Interpretative Introduction*, p. 36.
15. Richard Thurnwald, *Economics in Primitive Communities.* London: Oxford University Press, 1932, p. xiii.
16. Karl Polanyi, *Great Transformations.* Beacon Press, Boston, 1944, p. 47.
17. Marshall Sahlins, 'Tribal Economics' in George Dalton (ed), *Economic Development and Social Change.* Natural History Press, New York, 1971, p. 44.
18. Ibid., p. 45.
19. It means, the lands cleared, cultivated for a few years and abandoned, and covered with a fresh growth of jungle. In other words, a land in the hill areas capable of cultivation after long intervals with particular kinds of grain sown in holes dug with a spade.
20. Francis Buchanan, *A Journey from Madras through the Countries of Mysore, Canara and Malabar*, 1807, p. 198.
21. M.D. Subash Chandran, 'Shifting Cultivation, Sacred Groves and Conflicts in Colonial Forest Policy in Western Ghats' in Richard H Grove, Satpal Sangwan and Vinita Damodaran (eds) *Nature and the Orient: The*

Environmental History of South and Southeast Asia. Delhi: Oxford University Press, 1998, pp. 678–700; Eric Meyer, 'Forests, Chena Cultivation, Plantations and the Colonial State in Ceylon 1840–1940' in Richard H Grove, Satpal Sangwan and Vinita Damodaran (eds) *Nature and the Orient: The Environmental History of South and Southeast Asia*. Oxford University Press, Delhi, 1998, p. 797; Felix Padel, 'Forest Knowledge: Tribal People, their Environment and the Structure of Power' in Richard H Grove, Satpal Sangwan and Vinita Damodaran (eds.), *Nature and the Orient: The Environmental History of South and Southeast Asia*. Delhi: Oxford University Press, 1998, p. 899.

22. Guha and Gadgil, 'State Forestry and Social Conflict' p. 152; Jacques Pouchepadass, 'British Attitudes towards Shifting Cultivation in Colonial South India: A Case Study of South Canara District 1800–1920' in David Arnolad and Ramachandra Guha (eds), *Nature, Culture, Imperialism: Essays on the Environmental History of South Asia*, Delhi: Oxford University Press, 1995, p. 135.
23. Chandran, 'Shifting Cultivation, Sacred Groves and Conflicts', p. 688; Pouchepadass, 'British Attitudes towards Shifting Cultivation', p. 148.
24. Chandran, 'Shifting Cultivation, Sacred Groves and Conflicts', p. 675; Meyer, 'Forests, Chena Cultivation, Plantations and the Colonial State' p. 797.
25. S. Cox, 'The Forest of Madras' in Somerset Playne (complied) *South India*: Foreign and Colonial Compiling and Publishing & Co, London, 1914–1915.
26. Madhav Gadgil, 'Towards an Ecological History of India', *Economic and Political Weekly*, 20(45, 46 and 47), 1985, p. 1913.
27. John Shortt, *Hill Ranges of Southern India* (ed), Higginbotham and Co, Madras, 1870, p. 51.
28. Saravanan, Colonial Commercial Forest Policy and the Tribal Private Forests'; Saravanan, 'Colonialism and Coffee Plantations'; Saravanan, 'Commercial Crops, Alienation of Common Property Resources and Change in Tribal.
29. Velayutham Saravanan, 'Economic Transformation of Tribals in Tamil Nadu: Causes of Migration and its Employment Pattern, 1792–1991' *Indian Journal of Labour Economics*, 41(4), 1998, pp. 695–705; Saravanan, 'Colonial Commercial Forest Policy and the Tribal Private Forests'.
30. A. Sarada Raj, Economic Conditions in the Madras Presidency 1800–1850, Madras: University of Madras, 1941, p. 14; Saxena, Forest Policy in India.
31. Sarada Raj, *Economic Conditions in the Madras Presidency*, p. 151.
32. E. P. Stebbing, *The Forests of India*, Vol. I, London: John Lane The Bodley Head Limited, 1922, p. 34.

33. Stebbing, The Forests of India; H. Colyear Dawkins and Michael. S. Philip, *Tropical Moist Forest Silviculture and Management: A History of Success and Failure*, CAB International, Wallingford, UK, 1998.
34. Stebbing, The Forests of India, p. 64.
35. Ibid.
36. Ibid., p. 34.
37. Ibid., p. 64.
38. C.D. Maclean, *Manual of the Administration of the Madras Presidency*, Vol. 1, Madras: Government Press, 1885, p. 344.
39. Saravanan, 'Commercialisation of Forest, Environmental Negligence and Alienation of Tribal Rights'.
40. A measure of weight that prevailed in Madras Presidency: 1 *mound* is equal to 25 lbs in the Madras Presidency.
41. One *Star Pagoda* was equal to 45 *fanams* or 3600 cash; 1 *Star Pagoda* is equal to 3½ Company Rupees.
42. A money measurement, 80 *cash* is equal to 1 *fanam*; 45 *fanams* are equal to 1 *Star Pagoda*.
43. This was a lowest money measurement: 80 *cash* is equal to 1 *fanam* and 45 *fanams* are equal to 1 *Star Pagoda*.
44. Letter from Board of Revenue to the Collector of Salem, 6 July 1801 (Chennai: Tamil Nadu State Archives (hereafter TNSA).
45. A dry grain measurement: 1 *candies* is equal to 40 *bullas* or *puddies* or 7248 cubic inches.
46. Board of Revenue, Vol. 258, July 1800, p. 6642, TNSA.
47. BOR, Vol. 1587, 11 December, 1837 (Chennai: Tamil Nadu State Archives), p. 1504.
48. Maclean, *Manual of the Administration*, p. 290.
49. Saravanan, 'Colonialism and Coffee Plantations' Saravanan, 'Commercial Crops, Alienation of Common Property Resources'.
50. Maclean, *Manual of the Administration*, p. 336.
51. Saravanan, 'Colonialism and Coffee Plantations'; Saravanan, 'Commercial Crops, Alienation of Common Property Resources'.
52. Maclean, *Manual of the Administration*, p. 290.
53. Saravanan, 'Colonialism and Coffee Plantations'.
54. Saravanan, 'Commercial Crops, Alienation of Common Property Resources'.
55. Sarada Raj, *Economic Conditions in the Madras Presidency*, p. 150.
56. Ibid., p. 151.
57. Ibid., p. 155.
58. A.J. Arbuthnot, Sir Thomas Munro: Selections from Minutes and other Official Writings, London, 1881, p. 561.

59. Dietrich Brandis, Suggestions Regarding Forest Administration in the Madras Presidency. Madras: Government Press, 1883, p. 55.
60. Ibid.
61. Ibid.
62. Sarada Raj, *Economic Conditions in the Madras Presidency*, p. 157.
63. Ibid.
64. Ibid.
65. Brandis, *Suggestions Regarding Forest Administration*, p. 55.
66. Saravanan, 'Commercialisation of Forest, Environmental Negligence and Alienation of Tribal Rights'.
67. Ibid.
68. Brandis, *Suggestions Regarding Forest Administration*, p. 40.
69. F.J. Richards, *Madras District Gazetteers, Salem District*, Vol. I, Part I, Madras: Government Press, 1918, p. 248.
70. Saravanan, 'Colonial Commercial Forest Policy and the Tribal Private Forests'.
71. B.B. Chaudhury, 'Adivasi and Aranyaka: Reconsidering some characterizations of their polity and economy in pre–colonial and colonial India' in B.B. Chaudhuri and Arun Bandopadhyay (eds), *Tribes, Forest and Social Formation in Indian History*. New Delhi: Manohar, 2004, p. 95.
72. Francis Buchanan, *A Journey from Madras through the Countries of Mysore, Canara and Malabar*, Vol. 1, London: Directors of the East India Company, 1807, p. 198.
73. Saravanan, 'Commercialisation of Forest, Environmental Negligence and Alienation of Tribal Rights'.
74. Saravanan, 'Colonialism and Coffee Plantations'.
75. Saravanan, 'Commercialisation of Forest, Environmental Negligence and Alienation of Tribal Rights'.
76. Ibid.
77. Letter from Colonel R.H. Beddome, Conservator of Forests to J.H. Garstin, Acting Secretary to Government, revenue Department, dated 22 December 1876 (Chennai: Tamil Nadu State Archives).
78. Velayutham Saravanan, 'Agrarian Policies in the Tribal Areas of Madras Presidency During the Pre–Survey and Settlement Period, 1792–1872' *Indian Journal of Agricultural Economics*, 65(2), 2010, pp. 261–276.
79. Saravanan, 'Agrarian Policies in the Tribal Areas'.
80. Saravanan, 'Colonial Commercial Forest Policy and the Tribal Private Forests'; Saravanan, 'Colonialism and Coffee Plantations'.
81. As per the archival records, since early 1820s, coffee plantations were established mainly by the non-tribals. As a result, the settlement of non-tribals has been increased considerably in the hill regions.

82. 'In times prior to colonial rule the whole of the Pariah community, without exception, were the slaves of the superior castes' (Census of India 1871, p. 169).
83. Papers relating to the Coffee districts 1859.
84. John Shortt, *Hill Ranges of Southern India.* (ed.), Madras: Higginbotham and Co., 1870, p. 7.
85. H. Le Fanu, *A manual of the Salem District in the Presidency of Madras*, Vol. II, Madras: Government Press, 1883, pp. 21–24; Velayutham Saravanan, 'Tribal Land Alienation in Madras Presidency during the Colonial period: 1792–1947' *Review of Development and Change*, 6(1), 2000, pp. 73–104.
86. Saravanan, 'Tribal Land Alienation in Madras Presidency'.
87. Velayutham Saravanan, 'Tribal Revolts in India with reference to Salem and Baramahal Districts of Madras Presidency during the late 18th Century' *Artha Vijnana*, 41(1), 1999, pp. 67–81.
88. Board of Revenue, Vol. 1537, 10 November 1836 (Chennai: Tamil Nadu State Archives), pp. 18133–18136.
89. Saravanan, 'Colonialism and Coffee Plantations'.
90. Velayutham Saravanan, 'Colonialism and Conservation: Commercialisation of Forests and decline of tribals in Madras Presidency, 1882–1947' presented at the workshop on the 'Nature in the Raj: Colonial Environments and Indigenous Knowledge' organised by Centre for World Environmental History, Sussex University and the Centre for Studies in Science Policy, Jawaharlal Nehru University, New Delhi, August 18–19: 2006, pp. 1–48.
91. Government of India, The Scheduled Tribes (Recognition of Forest Rights) Bill, 2005, No.17014/4/2005-S&M(Pt.), Ministry of Tribal Affairs, 2005.
92. Ibid.
93. Velayutham Saravanan, 'Joint Forest Management in Andhra Pradesh: Who does the Offences?' *Asian Economic Review*, 48(2), 2006, pp. 231–238.
94. Saravanan, 'Colonialism and Coffee Plantations'.
95. Saravanan, 'Colonial Commercial Forest Policy and the Tribal Private Forests'; Saravanan, 'Colonialism and Coffee Plantations'.

Check for updates

CHAPTER 3

Conservation or Commercialisation, 1882–1947

Existing literature on environmental history is mostly confined to the period when the Forest Acts were introduced, particularly the 1870s and 1880s, with a few exceptions that address the pre-colonial period. How these statutes, which provided for the establishment of reserve forest, helped environmental protection has been well researched, as has the impact they had on the tribals' livelihood pattern. Here it must be mentioned that the available studies have not paid adequate attention to the post-Forest Act period to ascertain how forest resources were protected or plundered and the resultant impact on the livelihood sources of forest-dependent indigenous communities. Given this background, an attempt has been made in this chapter to analyse how the colonial government systematically commercialised forest resources even while laying stress on conservation by introducing reserve forests, and the impact of this on tribals in the Madras Presidency during the post-Forest Act phase of the colonial period (1882–1947).

In this context, the chapter raises an important question as to whether the colonial government succeeded in containing deforestation through conservation policies or whether it continued to commercialise forest resources by introducing restrictions on forest-dwellers and users in the Madras Presidency. Further, it discusses the extent to which the state

An earlier version of the chapter was published in *Journal of Forest Economics*, 17(4), 2011.

© The Author(s) 2018

V. Saravanan, *Environmental History and Tribals in Modern India*,
https://doi.org/10.1007/978-981-10-8052-4_3

exploited forest resources to meet the burgeoning demands of empire, trade and industry, and the consequent decline of tribals that occurred. Precisely, this is an attempt to understand colonial policies and their implications on environment and tribals in the Madras Presidency during the post-Forest Act phase of the British Raj (1882–1947).

Trends of Forest Areas in the Madras Presidency

Trends of Land Utilisation

Land utilisation trends have changed dramatically in the Madras Presidency during the post-Forest Act period. Available data clearly indicates an increase in forest cover from 89.81 lakh acres in 1884/1885 to 129.86 lakh acres in 1920/1921. It further increased to 132.07 lakh acres in the subsequent decade, before remaining more or less the same for the rest of the colonial period. In other words, the proportion of forest area witnessed a marginal increase from 15.57% of the total geographical area in 1884/1885 to 16.40% in 1946/1947. Land not available for cultivation rose from 111.01 lakh acres in 1884/1885 to 237.39 lakh acres in 1910/1911, but decreased in the subsequent decades, in 1946/1947 dropping to 140.54 lakh acres. Though cultivable waste other than fallow land increased, its proportion declined sharply during the post-Forest Act phase. Fallow land increased from 9.82% in 1884/1885 to 12% in 1946/1947 owing to famine, drought and other natural calamities during the period (Table 3.1). Even as vast tracts of land became unsuitable for cultivation, cultivable waste other than fallow were brought under the forest cover. In other words, the common property resources other than the forests slowly vanished during this period.

Expansion of Agriculture

The expansion of agriculture was notable in the Madras Presidency during the post-Forest Act period. Despite an increase in fallow land, the net sown area increased progressively during this period. For instance, in 1884/1885, 213.32 lakh acres of land was the net sown area and this shot up to 310.35 lakh acres in 1946–1947. In other words, the proportion of net sown area, which was 37% of the total geographical area in 1884/1885, had risen to 39.05% in 1946/1947. This is proof that more and more land was being brought under cultivation during the late nineteenth and early twentieth centuries.

Table 3.1 Trends of land utilisation in the Madras Presidency, 1882–1947 (area in 000 ha)

	1884–1885	*1890–1891*	*1900–1901*	*1910–1911*	*1920–1921*	*1930–1931*	*1940–1941*	*1946–1947*	*1884–1885*	*1890–1891*	*1900–1901*	*1910–1911*	*1920–1921*	*1930–1931*	*1940–1941*	*1946–1947*
Forests	8.981	11.542	12.386	13.607	12.986	13.207	13.189	13.497	15.57	19.16	19.16	15.31	14.46	14.51	16.52	16.40
Not available for cultivation	11.101	11.96	12.638	23.739	21.827	20.146	14.508	14.054	19.25	19.85	19.85	26.71	24.31	22.13	18.17	17.68
Cultivable waste other than fallow	10.601	7.869	6.260	9.319	11.997	13.126	11.097	11.849	18.38	13.06	13.06	10.49	13.36	14.42	13.90	14.91
Current fallow	5.663	5.176	5.763	8.453	9.878	10.326	9.572	9.499	9.82	8.59	8.59	9.51	11.00	11.34	11.99	11.95
Net area sown	21.332	23.702	24.509	33.752	33.105	34.224	31.472	31.035	36.99	39.34	39.34	37.98	36.87	37.60	39.42	39.05
Total geographical area	57.677	60.249	61.556	88.87	89.793	91.029	79.838	79.472	100.00	100.00	100.00	100.00	100.00	100.00	100.00	100.00

Source: *Madras Administration Report* and *Season and Crops Reports* (various years)

Trend of Forest Areas

In the Madras Presidency, the area under green cover swelled from 89.81 lakh acres in 1884/1885 to 115.42 lakh acres in 1890/1891, and further increased to 130.35 lakh acres in 1946/1947. It is evident that within one and half decades of the implementation of the Madras Forest Act most forests had been brought under the reserve forest (Table 3.1). The leap in the size and spread of forest cover, however, witnessed certain fluctuations, almost every year that cannot be brushed aside. Roughly, about 14,597 square miles of forest area were excluded in nearly half of the years during the sixty-two years of post-Forest Act colonial period. This picture does not necessarily reflect the actual inclusion or exclusion but only the net impact. The exclusion of reserve forest was mainly to accommodate the needs of villages in terms of communal privileges, the extension of cultivation and more importantly the establishment of plantations.[1] Lands were also assigned to Indian soldiers who had given distinguished service in the wars and in order to remove forest boundaries in close proximity of villages. This indicates that to some extent the interest of the people and their needs were entertained during the introduction of the reserve forest scheme in the Madras Presidency.

Outline of Forest Policies (1882–1947)

Though the first Forest Act, known as Indian Forest Act 1865, had come into force much earlier in the Indian subcontinent, its provisions were not extended to the Madras Presidency. Even the Indian Forest Act 1878 was not implemented in the Madras Presidency. Although some Forest Acts were enacted much earlier, the Madras Forest Act 1882 was the first of its kind that was relevant to the Madras Presidency, and it was only when this statute came into force in 1883 that the conservation of forest resources gained any importance in the region.

Commercial motives rather than conservation were the driving force behind these initiatives, although cloaked under the garb of environmental protection. Here, it would be appropriate to quote Brandis (1883): 'The advantages expected to be secured by forest conservancy in the Madras Presidency, as in other provinces of the Indian Empire, are of two classes, direct and indirect. The direct advantages are easily defined. They comprise the provision of permanent and sufficient supply of timber, wood and other forest produce for local use and for railways, iron smelting, towns and the trade generally. Regarding the indirect advantages opinions differ. The extent to which forests influence the climate, the under-ground and surface drainage, remains still to be proved ... but certain indirect advantages are obvious, and these are: that in a hilly country, forests protect the surface soil against erosion, and diminish the quantity of sand and silt carried down by rivers; that they help to fix moving sands, afford shelter against the winds, particularly against the scorching winds of the hot season; and that, during the dry season, well-stocked forests tend to increase the atmospheric moisture and dew in their immediate vicinity.'[2] This clearly indicates that the motive for conservation was commercialisation rather than conservation per se from an environmental point of view. In other words, the commercial sustainability of forest resources was portrayed in terms of environmental importance.

The concept of reserve forest, introduced under the Forest Act, in effect excluded and alienated the tribals and other forest-dwellers from their traditional settings. A large proportion of forest land was brought under this scheme at the close of the nineteenth century. In the end almost all the hill regions were brought into the reserve forest at this time, with Salem and Baramahal region being a case in point.[3] However, there was a push to implement this in the subsequent period. For instance, during the first decade of the twentieth century about 15.31% of the total geographical area was brought into the reserve forest, this increasing to 16.40% at the end of colonial rule (1946/1947). In addition to the introduction of reserve forest, several measures were initiated to conserve forest. Besides a tax on goat rearing and a grazing fee, the number of cattle was also restricted. Violation of rules and restrictions attracted heavy penalties. To protect forest wealth, the number of forest administrators and staff was increased.

During the second quarter of the twentieth century, reserve forest was classified into three classes: production, protection and *ryot*. The *ryot* forest was earmarked to meet the grazing and fuel requirements of the local population. In 1925, forest *panchayats* were introduced to manage the forests and powers were transferred to the Revenue Department. In 1925/1926, about 623 *panchayats* were managing 1269 square miles of forest, with the area under them increasing in subsequent years; but it must be noted that more than half of the *panchayats* remained ineffective.[4] However, they continued until the close of colonial rule and about 19,000 square miles of land was transferred to the *ryot* forest, controlled by the Revenue Department.[5] In addition to this, several concessions were made during the Congress Ministry in 1937–1939, such as a reduction by half of grazing fees and an extension of grazing areas, allowing the free collection of grass and fuel and a larger area for shifting cultivation.[6] Later, much of the *ryot* forests, having been transferred to the *panchayats*, were returned to the Forest Department either owing to the inability of a *panchayat* to manage effectively and/or because of degradation through overuse.[7]

In 1936, two forest divisions, Ganjam and Parlakimedi, were transferred to Orissa Province. In 1946, about 21,096 acres (32 square miles) of reserve forest was delimited from the forest.[8] In 1944, a draft bill for the control of privately owned forests was published, and in 1946 the Madras Preservation of Private Forest Act 1946 was enacted.[9] In the name of conservation, a large amount of forest land was brought into the reserve forest, and restrictions were imposed on the use of the forest resources in the Madras Presidency.

Forest Acts and Restrictions on the Tribals

Prior to the Madras Forest Act 1882, several restrictions were imposed on the use of forest resources, but they remained confined to certain hills/regions, particularly where commercial crops such as coffee and tea plantations had been established.[10] For instance, during the early nineteenth century when coffee cultivation was introduced in the Shervaroy hills of the Madras Presidency, the tribals and other local people were deprived of a very large proportion of their land. Of course, the objective behind these measures was not preservation but the exclusion of tribals and other forest-dwellers from access to these resources.[11]

Several restrictions were imposed on the tribals and other forest-dwellers under the Madras Forest Act 1882. Under this, the tribals could claim only a right of way, watercourses, pasture and forest produce (Section 10). These were allowed wholly or partially by the discretion of the Forest Department (Section 11). Further, they were intermittently granted with quantitative restrictions (Section 12). Fresh clearings for cultivation or any other purpose were also restricted (Section 7). If the tribals used the prohibited items, they were liable for imprisonment for a period of six months or a fine of about Rs.500 or both, besides compensation for damage (Section 21). Grazing and collection of forest produce on forest land was limited and liable for suspension (Section 22). Grazing cattle in the closed lands attracted imprisonment for a month and/or a penalty of Rs.200

(Section 28). Penalties were similar to those for breaching any rules made by the Forest Department (Section 36).

The Madras Forest Act of 1882, Section 41 states: 'When there is reason to believe that a forest offence has been committed in respect of any timber or forest produce, such timber or produce, together with all tools, ropes, chains, beats, vehicles and cattle used in committing any such offence, may be seized by a Forest Officer or Police Officer.' Section 51 reads: 'Any Forest Officer or Police Officer may, without orders from a Magistrate and without a warrant, arrest any person reasonably suspected of having been concerned in any forest offence punishable with imprisonment for one month or upwards, if such person refuses to give his name and residence or gives a name or residence which there is reason to believe to be false, or if there is reason to believe he will abscond.'

Restriction on Grazing

In addition to the restrictions, a large proportion of the forest was closed either completely or partially to grazing in different parts of the Madras Presidency. After the introduction of the Madras Forest Act, the area closed to grazing increased progressively until the first quarter of the twentieth century. For instance, in 1901/1902, 2219 square miles or 13.38% was closed for grazing either partially or wholly and this nearly doubled to 4036 square miles or 21.17% in 1922/1923.[12] However, in the subsequent period the proportion of forest land closed to grazing considerably declined. Prior to 1925, about one-fifth of the forest area was closed to grazing. Following a series of protests from the people, this was reduced to one-eighth in 1925, one-tenth in 1935 and 9% in 1942.[13]

Restrictions on Cattle

Besides closing vast tracts of forest area for grazing, curbs were placed on rearing cattle with a ceiling on their numbers; this affected domestic cattle wealth in the hill regions. Moreover, paid grazing was introduced and in 1900/1991, about 40 lakh animals were permitted via this scheme. This number progressively declined: 36 lakh in 1910/1911, 20 lakh in 1920/1921, 15 lakh in 1930/1931 and 14 lakh in 1937/1938.[14] This clearly indicates that although the proportion of forest land opened up for grazing increased, the number of animals was restricted over the period.

Forest Hill Village System

Pursuant to the Madras Forest Act 1882, the Forest Department placed a number of restrictions on the tribals. The forest hill village system introduced in different hills of the Madras Presidency during the early twentieth century is one such example. Under this, no goats were allowed; cattle of foreign origin and the people from plains were not admitted into the villages; villages were to protect the land allotted to their care, particularly protecting sandalwood trees from theft or other damage; the boundary lines and footpaths were to be kept tidy by the villagers; and they had to supply labour for all departmental operations in their respective areas and carry the luggage of touring officers, for which only a meagre payment was made. In case of any lapse in protection, the deposit collected from the village would be forfeited in whole or in part, according to the amount of damage. Outsiders promptly issued notices of contempt against the tribals for all forest offences committed. The tribals received free of charge only small timber, bamboo and firewood, besides tree trunks for agricultural works.[15]

Grazing Fees

The forest hill village system was implemented in the Chitteri hills during 1915/1916, in the Namakkal-Kolli hills in 1917 and in Aranuthumalai by 1920/1921.[16] Under this system in the Namakkal-Kolli hills, 3 *annas* per cow unit were collected for the protection of forests from fire, theft of forest produce and so on.[17] This was supposed to be returned at the end of the year, according to the degree of protection maintained, but usually the officers or their agents did not return it.

In 1919, the Conservator of Forests proposed an increase in the grazing fee to 6 *annas* in view of the extension of the grazing area. In 1921, this was again raised to 9 *annas* per cow, in order to maintain a uniform rate throughout the Salem and Baramahal district. The Collector who accepted the fee enhancement refused to offer the service to anyone other than the Malaiyalis.[18] The same practice was followed in the Chitteri and Aranuthumalai hills.[19]

This system turned the tribals' rights over the forest into concessions. In other words, they were permitted only to use the forest produce and to graze their cattle, and were held responsible for protecting the forests from theft, fire and so on. Owing to the high grazing fees and other conditions, the tribals were forced to minimise their dependence on forests.

From 1933, the number of cattle permitted for grazing was restricted in order to prevent the clandestine import of cattle from the plains. For any increase in the numbers above the prescribed limit, the Malaiyalis/tribals were to pay an extra amount, which would be reimbursed later.[20] In two reserves of the Kolli hills, Gundur and Ariyur, free grazing was permitted for a limited number of cows, 491 and 217 respectively.[21] In one of the two forest *panchayats* of Jarugumalai, the Malaiyalis were given a concession to graze their cattle at half the prescribed fee.[22] In the remaining reserves of the Kolli hills and Jarugumalai and in respect of all the reserves in Bodamalai, Pachamalai and Shervaroy hills, the Malaiyalis had to obtain permits for grazing at 9 *annas* per cow, and in Namakkal taluk the permit was 6 *annas*.[23] That this was too high a rate for the tribals was substantiated by the settlement officer: 'the inability of the Malaiyalis to pay the grazing fee formed the main item of representation and some relief may be necessary'.[24] The Forest Department also enforced many restrictions even on the limited rights permitted to the tribals, which resulted in reductions in their livelihood options. A great extent of the forests was brought under reserve forest rules, with the Forest Acts and further rules and regulations being imposed on forest-dwellers and users during the post-Forest Act colonial period.

Commercialisation of Forests

A wide range of forest resources continued to be exploited for commercial purposes after the introduction of forest reserves under the Madras Forest Act. The export of forest products to several countries had been on the rise since the late nineteenth century, with coffee and sandalwood the major items.[25] From the late eighteenth century, commercial exploitation was the guiding principle of colonial forest policy, and conservation was not a priority until the late nineteenth century, when the reserve forest concept was introduced.[26] However, the government did not entirely give up commercial motives until the first half of the twentieth century. Even the very idea of forest conservation seems to have been accepted in order to satisfy the commercial requirements for scarce forest resources. In 1923/1924, the Forest Department appointed a Chief Forest Engineer to evolve a new process of exploiting forest resources. Another officer, known as the Forest Utilisation Officer, was appointed to deal exclusively with the commercial side of the Forest Department.[27] During the post-Forest Act period, forest resources were either depleted or cleared for a variety of purposes, such as the expansion of railway networks, the introduction of plantations, the extraction of timber and fuel-wood, and the export of products such as sandalwood.

Expansion of Railways

The establishment of the railways in 1853 and their expansion in subsequent years depleted the forest resources extensively in different parts of the country, both for laying the railway track and making berths.[28] Further, the expansion of the railway system led to the exploitation of forests, which adversely affected the physical environment as well as the livelihood of forest-dwellers in different parts of the country, having a grave impact on them.[29] However, in the first three decades of their existence, the expansion of the railways was limited to less than 6000 miles. In the succeeding years, the length of the network increased remarkably. By 1883 it had shot up to 10,198 miles (16,412 kilometres) and by the 1940s to over 40,000 miles (65,218 kilometres).[30] This growth was made possible by the felling of a very large number of trees and the clearing of vast tracts of jungles during the post-Forest Act period.

The adverse effects could be broadly divided into two categories: direct and indirect. The first pertained to felling and clearing as well as to make berth in the train coach. The second related to the transport of timber on a massive scale on the railways, and there was a steady growth in this from the late nineteenth century, from 464,900 tons in 1883 to 2,287,500 tons in 1946/1947. In total, about 85,230,985 tons of timber was transported through the railways between 1882 and 1946/1947.[31] The large quantity of timber being transported indicates a clear exploitation of forest resources from the late nineteenth century.

Expansion of Railways in the Madras Presidency

In the Madras Presidency, the first railway line was opened on 1 July 1856 by the Madras Railway Company: it ran for 63 miles between Veyasarpandy and Walajah Road (Arcot). The network had increased to 690 miles by 1866 and to 1436 miles by 1877.[32] Prior to the Madras Forest Act 1882, about 1523 miles of railway track had been laid in the Madras Presidency and other native states. Until the first decade of the twentieth century, the quantity of railway track increased remarkably: in 1882, in the Madras Presidency it was about 1523 miles and it increased to 4536 miles in 1907.

The growth rate was about 195% in 1883–1907. By 1920, the amount of track had increased marginally to 4687 miles.[33] Between 1907 and 1920, the growth rate of the railway network was therefore only about 4.37%.

Besides felling trees to allow for the laying of railway track, forests were denuded to supply fuel-wood to the railways as well as to other industries. The consumption of fuel-wood by the railways witnessed a leap from 9821 tons in 1863 to 54,358 tons in 1868.[34] Between 1873 and 1877, about 354,921 tons of wood were purchased by the railways for fuel. The fuel-wood requirement did not decrease in the years that followed. In 1878, the Madras Railways agent stated that the approximate requirement of wood for fuel was about 91,000 tons.[35] An increase in the length of railway network, number of berths in the railway coach, and fuel-wood all led to an increase in the exploitation of forest resources.

Timber for Sleepers

It is clear that from the second half of the nineteenth century a large quantity of timber was cut down for the railways in different parts of the Madras Presidency.[36] This continued during the post-Forest Act period. For instance, about 807,545 sleepers were supplied to the railways over the period of fifteen years between 1925/1926 and 1940/1941 (Table 3.2). Nearly half of them consisted of broad gauge sleepers. This indicates that several thousands of trees were cut down in the Madras Presidency during the post-Forest Act period.

Sandalwood

Sandalwood trees were largely found in the hilly tracts of North Arcot, Salem, Dharmapuri and Coimbatore districts, while the Nilgiris, South Arcot, Tiruchirapally, Ramanathapuram and Tirunelveli districts were not endowed with large quantities of this precious species. From the late eighteenth century, sandalwood was extracted for commercial purposes and even exported from various parts of the Madras Presidency.[37] In fact, sandalwood was a major source of revenue for the Forest Department.

A large quantity of sandalwood was extracted during the post-Forest Act colonial period, this progressively increasing (Table 3.3), with about 22,755 tons being extracted between 1882 and 1947. This clearly indicates that the extraction of forest resources did not decline when the concept of conservation was introduced.

Table 3.2 Number of sleepers supplied for railways, 1925/1926–1940/1941

Year	*No. of broad gauge sleepers*	*No. of metre gauge sleepers*	*Total no. of sleepers*
1925/1926	13,451	13,593	27,044
1926/1927	17,735	35,951	53,686
1927/1928	21,306	13,138	34,444
1928/1929	37,646	18,048	55,694
1929/1930	41,234	13,733	54,967
1930/1931	35,000	52,600	87,600
1931/1932	35,764	29,512	65,276
1932/1933	N.A	N.A	66,262
1933/1934	23,475	33,362	56,837
1934/1935	1702	26,501	28,203
1935/1936	30,270	36,472	66,742
1936/1937	29,776	40,014	69,790
1938/1939	30,000	17,000	47,000
1939/1940	30,000	17,000	47,000
1940/1941	30,000	17,000	47,000
Total	377,359	363,924	807,545

Sources: *Madras Administration Report* and *Administration Report of the Forest Department of the Madras Presidency* (various years)

N.A not available

The price of sandalwood increased but fluctuated every year during the post-Forest Act period. For example, in the Madras Presidency, the cost of 1 ton was about Rs.365 in 1900/1901, increasing to Rs.1600 in 1920/1921. Though the rate declined to Rs.782 in 1930/1931 and Rs.605 in 1940/1941, it increased to Rs.1956 in 1947.[38] If the price of sandalwood had not been as high as in the post-independence period, it would not have attracted so much attention from smugglers and illicit felling.[39] This indicates that the exploitation of sandalwood continued, despite the measures aimed at conservation.

Coffee and Tea Plantations

Coffee and tea plantations were established in different hill regions of south India during the early nineteenth century, particularly between the third and fifth decades.[40] Coffee plantations were largely found in the highlands of the west coast regions: Mysore, Coorg, Travancore, Wynad and the Nilgiris during the nineteenth century.[41] Apart from Wynad and

Table 3.3 Quantity and value of sandalwood extracted from the forests in the Madras Presidency, 1882/1883–1946/1947

Years	*Sandal-wood (quantity in tons)*	*Sandal-wood (Rs. in lakhs)*	*Year*	*Sandal-wood (quantity in tons)*	*Sandal-wood (Rs. in lakhs)*	*Year*	*Sandal-wood (quantity in tons)*	*Sandal-wood (Rs. in lakhs)*	*Year*	*Sandal-wood (quantity in tons)*	*Sandal-wood (Rs. in lakhs)*
1882/1883	21	N.A	1898/1899	N.A	0.28	1914/1915	607	4.76	1930/1931	564	4.41
1883/1884	N.A	N.A	1899/1900	64	N.A	1915/1916	N.A	N.A	1931/1932	672	6.13
1884/1885	N.A	0.16	1900/1901	106	N.A	1916/1917	N.A	N.A	1932/1933	648	5.95
1885/1886	64	N.A	1901/1902	76	N.A	1917/1918	N.A	5.24	1933/1934	511	4.58
1886/1887	11	N.A	1902/1903	69	N.A	1918/1919	432	6.18	1934/1935	604	4.23
1887/1888	18	N.A	1903/1904	79	N.A	1919/1920	497	7.16	1935/1936	716	4.92
1888/1889	90	0.11	1904/1905	116	N.A	1920/1921	418	6.68	1936/1937	960	5.52
1889/1990	114	N.A	1905/1906	153	N.A	1921/1922	521	6.05	1937/1938	1248	7.67
1890/1991	74	0.14	1906/1907	148	N.A	1922/1923	468	5.54	1938/1939	1204	6.86
1891/1992	97	N.A	1907/1908	164	N.A	1923/1924	710	7.86	1939/1940	1063	6.44
1892/1993	71	N.A	1908/1909	496	N.A	1924/1925	703	8.88	1940/1941	N.A	N.A
1893/1994	70	N.A	1909/1910	298	N.A	1925/1926	641	7.53	1941/1942	N.A	N.A
1894/1995	60	0.42	1910/1911	274	N.A	1926/1927	724	9.28	1942/1943	N.A	N.A
1895/1996	54	0.21	1911/1912	334	N.A	1927/1928	677	9.24	1943/1944	1233	14.70
1896/1997	N.A	N.A	1912/1913	293	N.A	1928/1929	579	8.60	1944/1945	1023	20.83
1897/1998	N.A	N.A	1913/1914	273	1.52	1929/1930	707	9.87	1945/1946	938	

Source: E.S. Thangam, 'Forest Products of Madras State' in *Forest Centenary of Forest Administration in Madras State*, Madras Forest Department, 1959, p. 21 and *Madras Administration Reports* (for various years)

N.A not available

the Nilgiris in the Madras Presidency, coffee had been cultivated in the Shervaroy hills of Salem and Baramahal regions since the early nineteenth century. In 1884/1885, about 60,933 acres of land was under coffee cultivation, and this had increased to 64,181 acres by 1900/1901.[42]

Unlike the coffee plantations, tea estates were found in Vizagapatanam, Madurai, the Nilgiris and Malabar districts of the Madras Presidency and Travancore Princely States. In 1883, there were about eighty-six estates with 11,874.75 acres in the Madras Presidency and twenty-six estates with 1005 acres in Travancore States.[43] Coffee plantations, although established during the pre-Forest Act period, grew in size during the post-Forest Act colonial period. For instance, in Salem and Baramahal regions alone the extent of coffee plantations was nearly 6224 acres in 1900 and increased to 8257 acres in 1932/1933.[44] The geographical spread of plantation crops indicates that a vast amount of forest land was diverted for the cultivation of commercial crops.

The post-Forest Act period saw a rapid increase of the area under plantation in different parts of the Madras Presidency. For instance, in 1913/1914, about 70,344 acres (coffee 43,522 acres and tea 26,822 acres) was under plantation crops,[45] and this rose sharply to about 116,986 acres (coffee 51,377 acres and tea 65,609 acres) in 1930/1931. The growth of the plantation sector necessarily meant denudation of a large amount of green cover.

Export of Coffee and Tea

A vast expanse of dense jungle was brought under the cultivation of commercial crops, and a large quantity of the produce was exported to other countries during the post-Forest Act period. Coffee was exported predominantly to the United Kingdom and France as well as Austria, Turkey, Egypt, the United States of America, Arabia, Ceylon, Persia, Victoria, foreign ports in India and other ports of the Indian subcontinent.[46] During the post-Forest Act period, coffee and tea were exported to different countries, its quantity increasing progressively over the period (Table 3.4). The data clearly indicates that the demand for coffee and tea both within the country and outside resulted in an increase in cultivation, which substantially affected forest resources during the post-Forest Act period.

Table 3.4 Export of coffee and tea from the Madras Presidency, 1882/1883–1946/1947 (Rs. in lakhs)

Year	*Coffee*	*Tea*	*Year*	*Coffee*	*Tea*	*Year*	*Coffee*	*Tea*	*Year*	*Coffee*	*Tea*
1882/1883	135.00		1898/1899	176.06	28.21	1914/1915	157.96	145.14	1930/1931	188.00	412.00
1883/1884	145.51	2.55	1899/1900	95.08	32.63	1915/1916	92.12	173.76	1931/1932	93.00	391.00
1884/1885	125.28	3.14	1900/1901	124.06	32.59	1916/1917	75.36	175.85	1932/1933	108.88	460.59
1885/1886	135.00	N.A	1901/1902	127.39	33.70	1917/1918	43.43	157.18	1933/1934	101.61	399.40
1886/1887	151.62	3.81	1902/1903	131.47	N.A	1918/1919	83.40	N.A	1934/1935	72.50	425.73
1887/1888	153.09	4.19	1903/1904	N.A	N.A	1919/1920	146.72	N.A	1935/1936	101.23	441.62
1888/1889	185.75	4.79	1904/1905	165.66	N.A	1920/1921	N.A	N.A	1936/1937	82.02	N.A
1889/1890	150.68	4.88	1905/1906	171.61	N.A	1921/1922	N.A	N.A	1937/1938	51.90	453.26
1890/1891	148.74	8.13	1906/1907	96.91	N.A	1922/1923	N.A	N.A	1938/1939	72.02	491.59
1891/1892	N.A	N.A	1907/1908	110.48	90.46	1923/1924	32.96	286.99	1939/1940	79.00	540.27
1892/1893	N.A	N.A	1908/1909	138.00	112.00	1924/1925	37.24	301.03	1940/1941	N.A	N.A
1893/1894	198.79	14.96	1909/1910	108.85	115.73	1925/1926	24.49	353.38	1941/1942	N.A	N.A
1894/1895	213.75	18.66	1910/1911	131.24	122.89	1926/1927	49.74	363.72	1942/1943	N.A	N.A
1895/1896	224.90	20.76	1911/1912	130.75	123.58	1927/1928	92.52	391.17	1943/1944	51.81	691.41
1896/1897	N.A	N.A	1912/1913	153.88	136.01	1928/1929	N.A	419.08	1944/1945	22.87	632.36
1897/1898	154.79	25.78	1913/1914	152.41	146.16	1929/1930	142.00	425.00	1945/1946	N.A	625.30

Source: *Madras Administration Reports* (for various years)

N.A not available

Exploitation of Bamboo

In addition to bringing more forest area under commercial crops, a large number of bamboo trees were cut down for commercial purposes during the post-Forest Act period. For example, in 1888/1989 about 111 lakh bamboo trees were felled, but this number shot up to over 745 lakhs in 1912/1913. In total, about 9230 lakh bamboos were cut down over a period of twenty-six years in the late nineteenth and early twentieth centuries (1888/1889–1913/1914 (Table 3.5). Though the data pertaining to felling is not available for the subsequent period, financial estimates about the trees cut down are available. The amount realised from bamboo increased progressively from Rs.2.46 lakh in 1914/1915 to Rs.13.67 lakh in 1946/1947. In other words, bamboo trees worth about 137.07 lakh were chopped down during the period 1914/1915–1946/1947 (Table 3.5).

Exploitation of Timber and Fuel Wood

From the late eighteenth century, forest resources were exploited for commercial purposes in different parts of the Madras Presidency.[47] In fact, the same kind of commercial forest policy continued during the post-Forest Act period, when conservation was emphasised. Though the Madras Forest Act imposed several restrictions on the tribals and other forest users and extended the area under reserve forests, the commercialisation of forest resources was one of the major concerns of the Forest Department. Indeed, it progressively increased during the post-Forest Act period. About 94,977 tons of timber and firewood was extracted in 1882/1883 and this increased to 777,960 tons in 1946/1947 (Table 3.6). In other words, about 68,669,567 tons (10,572,708 tons of timber and 58,096,859 tons of firewood) of forest resources were exploited between 1882/1883 and 1946/1947. The growth rate of timber and firewood exploitation was about 719.10% between 1882/1883 and 1946/1947. Exploitation of forest resources in general was very high particularly during the Second World War.

Export of Timber

After the introduction of the Madras Forest Act, a large quantity of timber was exported to other countries. Rosewood, adina and terminaha tomentos formed the bulk of exports. In 1920, timber was sent to London for display at the British Empire Timber Exhibition. In 1927/1928, the gov-

Table 3.5 Number of bamboo trees cut down and their value in the Madras Presidency, 1888/1889–1946/1947

Year	*No. of bamboo trees*	*Year*	*No. of bamboo trees*	*Year*	*Value (Rs. in lakh)*	*Year*	*Value (Rs. in lakh)*
1888/1889	11,100,840	1903/1904	37,025,554	1914/1915	2.46	1929/1930	4.55
1889/1890	12,694,181	1904/1905	35,826,887	1915/1916	3.17	1930/1931	3.94
1890/1891	18,223,447	1905/1906	32,083,686	1916/1917	3.43	1931/1932	3.44
1891/1892	29,793,195	1906/1907	39,216,767	1917/1918	3.25	1932/1933	3.83
1892/1893	33,725,819	1907/1908	47,048,184	1918/1919	3.16	1933/1934	3.88
1893/1894	30,846,684	1908/1909	32,032,765	1919/1920	3.14	1934/1935	3.95
1894/1895	29,317,049	1909/1910	39,419,784	1920/1921	2.64	1935/1936	3.90
1895/1896	26,081,017	1910/1911	53,000,000	1921/1922	3.24	1936/1937	3.74
1896/1897	29,958,554	1911/1912	47,048,184	1922/1923	4.08	1937/1938	3.93
1897/1898	39,217,618	1912/1913	74,553,732	1923/1924	4.03	1938/1939	3.97
1898/1899	35,844,728	1913/1914	41,000,000	1924/1925	3.45	1939/1940	3.68
1899/1900	37,025,554			1925/1926	4.13	1943/1944	7.36
1900/1901	35,826,887			1926/1927	4.11	1944/1945	10.97
1901/1902	39,217,618			1927/1928	3.52	1945/1946	10.40
1902/1903	35,844,728			1928/1929	4.05	1946/1947	13.67

Source: *Madras Administration Report* (for various years)

Note: Data is not available for the years of 1940/1941, 1941/1942 and 1942/1943

Table 3.6 Yield of forest products in the Madras Presidency, 1882/1883–1946/1947 (in tons)

Years	*Timber*	*Firewood*	*Total*	*Year*	*Timber*	*Firewood*	*Total*
1882/1883	55,338	39,639	94,977	1915/1916	78,380	397,240	475,620
1883/1884	11,773	15,596	27,369	1916/1917	85,720	420,940	506,660
1884/1885	22,828	23,548	46,376	1917/1918	84,860	460,440	545,300
1885/1886	19,573	40,136	59,709	1918/1919	79,080	371,000	450,080
1886/1887	42,042	25,426	67,468	1919/1920	84,640	239,760	324,400
1887/1888	13,383	24,382	37,765	1920/1921	78,760	391,460	470,220
1888/1889	63,295	261,968	325,263	1921/1922	86,400	389,600	476,000
1889/1890	6878	249,736	256,614	1922/1923	104,800	394,260	499,060
1890/1891	74,645	242,326	316,971	1923/1924	99,520	415,480	515,000
1891/1892	58,757	279,392	338,149	1924/1925	142,980	447,640	590,620
1892/1893	57,483	241,807	299,290	1925/1926	126,000	486,080	612,080
1893/1894	59,547	242,036	301,583	1926/1927	94,140	418,640	512,780
1894/1895	53,240	298,921	352,161	1927/1928	120,680	460,840	581,520
1895/1896	24,305	274,412	298,717	1928/1929	99,660	352,000	451,660
1896/1897	31,307	237,398	268,705	1929/1930	94,900	482,740	577,640
1897/1898	51,348	236,864	288,212	1930/1931	108,840	495,280	604,120
1898/1899	52,278	265,471	317,749	1931/1932	81,360	440,560	521,920
1899/1900	56,525	291,137	347,662	1932/1933	70,020	457,680	527,700
1900/1901	62,831	294,933	357,764	1933/1934	73,540	436,260	509,800
1901/1902	60,915	309,560	370,475	1934/1935	72,540	413,980	486,520
1902/1903	64,555	373,211	437,766	1935/1936	89,780	446,120	535,900
1903/1904	58,961	347,459	406,420	1936/1937	81,060	406,720	487,780
1904/1905	61,912	416,235	478,147	1937/1938	74,260	426,880	501,140
1905/1906	50,153	399,988	450,141	1938/1939	68,020	437,420	505,440
1906/1907	51,237	432,120	483,357	1939/1940	62,240	425,880	488,120
1907/1908	72,053	452,163	524,216	1940/1941	68,540	455,200	523,740
1908/1909	66,990	488,311	555,301	1941/1942	N.A	N.A	N.A

1909/1910	72,638	429,555	502,193	1942/1943	N.A	N.A	N.A
1910/1911	60,200	503,481	563,681	1943/1944	104,240	1,124,140	1,228,380
1911/1912	79,443	482,960	562,403	1944/1945	87,840	861,660	948,620
1912/1913	95,552	517,543	613,095	1945/1946	95,120	857,760	952,880
1913/1914	114,214	570,025	684,239	1946/1947	64,240	713,720	777,960
1914/1915	97,682	438,768	536,450				

Source: E.S. Thangam, 'Forest Products of Madras State' in *Forest Centenary of Forest Administration in Madras State*, Madras Forest Department, 1959, p. 21

N.A not available

ernment decided to appoint agents in various parts of the world to handle the timber business.[48] Timber export was undertaken both through agents, Messrs. Peirce, Leslie & Co and Messrs. Howard, Bros & Co., and Forest Utilisation Officers. In 1929/1930, about 282.33 tons of timber were exported to other countries, of which 125.50 tons were through the agents and 147.85 tons through Forest Utilisation Officers. Between 1883/1884 and 1916/1917, covering a period of 23 years, about Rs.374.40 lakh worth of timber was exported (Table 3.7). During the second and third decade of the twentieth century, covering a period of eleven years, about Rs.1488.71 lakh worth of timber was exported.

Revenue from Minor Forest Produce and Grazing Fees

In addition to the extraction of a large quantity of timber and fuel-wood, huge revenue was realised through minor forest produce and grazing fees. Overall, revenue from minor forest produce increased in spite of a decline

Table 3.7 Timber export from the Madras Presidency, 1882/1883–1946/1947 (Rs. in lakhs)

Year	*Value*	*Year*	*Value*	*Year*	*Value (in tons)*
1883/1884	18.45	1900/1901	20.58	1917/1918	N.A
1884/1885	14.66	1901/1902	15.80	1918/1919	N.A
1885/1886	N.A	1902/1903	N.A	1919/1920	N.A
1886/1887	18.17	1903/1904	N.A	1920/1921	100
1887/1888	16.36	1904/1905	N.A	1921/1922	N.A
1888/1889	16.29	1905/1906	N.A	1922/1923	N.A
1889/1890	19.00	1906/1907	N.A	1923/1924	12.73
1890/1891	19.30	1907/1908	N.A	1924/1925	246.78
1891/1892	N.A	1908/1909	N.A	1925/1926	223.83
1892/1893	N.A	1909/1910	10.48	1926/1927	231.63
1893/1894	19.97	1910/1911	17.33	1927/1928	16.8
1894/1895	17.11	1911/1912	18.68	1928/1929	102.25
1895/1896	18.22	1912/1913	20.99	1929/1930	282.33
1896/1897	N.A	1913/1914	23.35	1930/1931	136.25
1897/1898	16.06	1914–1915	7.29	1931/1932	87.93
1898/1899	13.81	1915/1916	14.35	1932/1933	Nil
1899/1900	2.55	1916/1917	15.60	1933/1934	48.18

Source: *Madras Administration Reports* and *Administration Report of the Forest Department of the Madras Presidency* (various years)

N.A not available

between 1924/1925 and 1946/1947. Revenue from grazing fees also declined progressively during the same period (Table 3.8). The opening up of vast tracts of forest land to grazing from the early twentieth century

Table 3.8 Revenue from minor forest produce and grazing fees, 1924/1925–1946/1947 (Rs. in lakhs)

Year	*Minor forest produce*	*Grazing fees*	*Total*	*Year*	*Minor forest produce*	*Grazing fees*	*Total*
1886/1887	N.A	3.57	3.57	1917/1918	N.A	7.40	7.40
1887/1888	N.A	1.06	1.06	1918/1919	N.A	8.24	8.24
1888/1889	4.55	1.30	5.85	1919/1920	N.A	8.53	8.53
1889/1890	4.58	1.43	6.01	1920/1921	N.A	8.17	8.17
1890/1891	5.81	1.50	7.31	1921/1922	N.A	7.90	7.90
1891/1892	N.A	N.A	N.A	1922/1923	N.A	9.24	9.24
1892/1893	N.A	N.A	N.A	1923/1924	N.A	N.A	N.A
1893/1894	N.A	2.67	2.67	1924/1925	8.03	7.53	15.56
1894/1895	8.81	3.28	12.09	1925/1926	8.03	7.53	15.56
1895/1896	10.80	3.40	14.20	1926/1927	8.03	7.53	15.56
1896/1897	9.43	3.76	13.19	1927/1928	8.03	7.53	15.56
1897/1898	9.99	3.64	13.63	1928/1929	8.83	7.53	16.36
1898/1899	9.87	4.34	14.21	1929/1930	9.03	6.21	15.24
1899/1900	10.92	4.90	15.82	1930/1931	6.44	6.21	12.65
1900/1901	11.91	5.10	17.01	1931/1932	6.03	5.59	11.62
1901/1902	11.98	5.40	17.38	1932/1933	5.67	5.86	11.53
1902/1903	11.44	5.79	17.23	1933/1934	5.55	5.84	11.39
1903/1904	12.67	6.07	18.74	1934/1935	5.42	6.02	11.44
1904/1905	13.13	6.27	19.40	1935/1936	5.44	5.90	11.34
1905/1906	14.03	6.63	20.66	1936/1937	5.68	5.97	11.65
1906/1907	14.15	6.79	20.94	1937/1938	5.87	6.00	11.87
1907/1908	15.00	7.27	22.27	1938/1939	5.29	3.80	9.09
1908/1909	15.80	7.83	23.63	1939/1940	5.14	3.87	9.01
1909/1910	16.93	7.70	24.63	1940/1941	5.86	4.15	10.01
1910/1911	14.11	6.32	20.43	1941/1942	2.83	4.03	6.86
1911/1912	15.70	6.32	22.02	1942/1943	N.A	N.A	N.A
1912/1913	N.A	6.71	6.71	1943/1944	5.31	4.20	9.51
1913/1914	16.00	6.82	22.82	1944/1945	8.21	4.11	12.32
1914/1915	14.64	6.96	21.60	1945/1946	9.22	6.60	15.82
1915/1916	N.A	6.87	6.87	1946/1947	10.9	7.19	18.09
1916/1917	N.A	7.04	7.04				

Source: *Madras Administration Report* (for various years)

N.A not available

onwards was probably the reason for this, and it should be seen in the context of about a quarter of the forest being closed for grazing in the first quarter of the twentieth century. However, in 1917/1918 the forest policy envisaged the abandonment of small and isolated reserves, which were expected to be taken over by the *panchayats*. To force this to happen, the government hiked the grazing fee in those areas. Hence, in the subsequent period the grazing fee was on the rise. In the second quarter of the twentieth century, the proportion of the area closed to grazing had declined progressively. In 1925, for example, about one-eighth of the forest area was closed to grazing, this being further reduced to one-tenth in 1935 and 9% in 1942.[49] In other words, more and more forest areas were opened up to grazing following representation and protest in different parts of the Madras Presidency. In addition to this, the grazing fee was reduced by 50% in 1938.

Second World War and Timber Exploitation

During the Second World War, particularly in early 1942, the fall of Burma and Andamans to the Japanese created an increased demand for timber from the Madras Presidency to meet the requirements of military and naval forces.[50] A large quantity of forest resources of different kinds was extracted at an alarming rate, forcing the government to restrict timber movement both by rail and sea, and merchants were asked to get permission from the Forest Utilisation Officer. Between 1939/1940 and 1945/1946, a large quantity of different kinds of forest resources were supplied to the military from the Madras Presidency. The details are as follows: in 1939/1940, about 4500 tons of timber, 120,000 bamboo *tatties* and 5400 tent poles; in 1940/1941, 27,990 tons of timber and 160,000 square feet of plywood boards; and in 1941/1942 about 21,300 tons of timber.[51] In 1942/1943, about Rs.67 lakhs worth of resources was supplied to the army from the Madras Presidency, this increasing to Rs.132 lakhs in 1943/1944, to Rs.178 lakhs in 1944/1945 and to Rs.133 lakhs in 1945/1946 (Table 3.9). Undeniably, the environmental degradation of the Madras Presidency could largely be attributed to the catastrophic effects of the war.

Table 3.9 Timber supply to army by the Madras forest utilisation division and timber supply circle, 1939/1940–1945/1946

Commodity	*1942/1943*		*1943/1944*		*1944/1945*		*1945/1946*	
	Tons	*Value*	*Tons*	*Value*	*Tons*	*Value*	*Tons*	*Value*
Teak logs	11,761	1,366,231	14,695	2,148,907	13,795	2,086,433	6622	1,039,469
Rosewood logs	–	–	661	103,897	1031	185,371	241	48,357
Hardwood logs	1650	139,815	22,598	1,707,516	30,914	2,815,586	26,699	2,490,315
Sleepers and Block nos	16,534		–	527,415	264,591	1,653,945	103,694	705,965
Sawn timber	33,905	1,282,144	65,319	7,845,928	69,410	7,794,566	66,452	7,688,645
Ballis and telegraph pole nos	70,741	3,820,178	552,981	486,078	–	1,231,722	–	1,206,544
Bamboos nos	–	62,000	819,486	163,556	–	1,567,224	–	3322
Bombax mats, sq. ft	–	–	8,922,060	155,755	–	–	–	–
Tent poles nos	–	–	8659	3681	–	66,666	–	–
Jetty piles	–	–	–	–	–	364,411	–	90,054
Total		6,670,368		13,151,733		17,765,924		13,272,671

Source: E.P. Stebbing, *The Forests of India*, 1962, p. 187

Trends of Income and Expenditure

The trends of the Forest Department's income and expenditure of offer a macro-picture of whether the colonial government's initiatives on conservation were guided by environmental concerns or commercial interests during the post-Forest Act colonial period. Apart from a substantial rise in revenue and expenditure, the Forest Department netted a huge income. For instance, revenue was about Rs.903,914 lakhs in 1882/1883, rising to Rs.15,867,746 lakhs in 1946/1947; expenditure increased from Rs.636,558 lakhs to Rs.7,971,742 lakhs in the same period; and Net income also increased from Rs.267,356 lakhs to Rs.7,896,004 lakhs. In other words, the revenue growth rate was about 1655.45% and expenditure was about 1152.32% between the period of 1882/1883 and 1946/1947. The growth rate of surplus revenue was about 2853.367% in the same period (Table 3.10). The above data clearly indicates that the Forest Department earned more revenue from the forest than the expenditure incurred. Except for during the great depression of the 1930s, timber output increased steadily, with a spurt created by the Second World War—when the government issued a direction to provide supplies to the military at a lower price. In summary, the Forest Department extracted more revenue from the forest than was spent, which clearly indicates that the commercialisation of forest resources was predominant over conservation in the Madras Presidency during the post-Forest Act period.

Decline of Tribal Livelihood Pattern

It is a well-known fact that the tribals' livelihood was interwoven with the forest and its resources. It has also been established beyond doubt that this livelihood witnessed an irreversible decline owing to the intervention of external forces from the early nineteenth century, particularly where plantations were established.[52] This decline was invariably visible throughout the Madras Presidency after the introduction of the Madras Forest Act of 1882. Restrictions focused on the use of forest resources, with the introduction of reserve forests, fees for grazing and the collection of other forest produce, which ultimately left the forest-oriented tribal economy in disarray during the post-Forest Act period. The powers to impose penalties that were vested with the law enforcement agencies led to severe harassment of the hapless tribals. In addition to this, the number of non-tribal settlements increased in the Madras Presidency's different hills. The

Table 3.10 Trends of revenue and expenditure of the Forest Department in the Madras Presidency, 1882/1883–1946/1947 (Rs. in lakhs)

Year	*Revenue*	*Expenditure*	*Net income*	*Year*	*Revenue*	*Expenditure*	*Net income*
1882/1883	903,914	636,558	267,356	1915/1916	3,938,475	2,607,949	1,330,526
1883/1884	958,991	480,916	478,075	1916/1917	4,779,106	2,784,374	1,994,732
1884/1885	1,162,689	884,421	278,268	1917/1918	5,416,496	2,919,596	2,496,900
1885/1886	1,198,600	936,431	262,169	1918/1919	4,916,509	3,131,939	1,784,570
1886/1887	1,246,783	1,155,687	91,096	1919/1920	5,953,910	3,516,428	2,437,482
1887/1888	1,374,920	1,128,517	246,403	1920/1921	4,616,184	3,437,522	1,178,662
1888/1889	1,515,006	1,216,674	298,332	1921/1922	4,796,900	4,842,366	−45,466
1889/1890	1,557,627	1,165,358	392,269	1922/1923	5,272,957	4,789,463	483,494
1890/1891	1,795,408	1,253,284	542,124	1923/1924	5,486,963	4,629,514	857,449
1891/1892	1,694,215	1,283,929	410,286	1924/1925	5,569,681	5,117,796	451,885
1892/1893	1,677,212	1,330,258	346,954	1925/1926	5,170,521	4,815,499	355,022
1893/1894	1,943,715	1,368,411	575,304	1926/1927	5,529,926	5,057,602	472,324
1894/1895	1,977,182	1,454,298	522,884	1927/1928	5,128,434	4,912,604	215,830
1895/1896	2,167,630	1,468,988	698,642	1928/1929	5,894,257	4,584,549	1,309,708
1896/1897	2,188,917	1,587,877	601,040	1929/1930	5,994,064	4,523,563	1,470,501
1897/1898	2,151,114	1,588,489	562,625	1930/1931	4,683,394	4,279,831	403,563
1898/1899	2,075,254	1,592,149	483,105	1931/1932	4,254,874	3,986,787	268,087
1899/1900	2,313,507	1,688,196	625,311	1932/1933	4,738,169	3,789,150	949,019
1900/1901	2,443,773	1,738,190	705,583	1933/1934	3,943,068	3,920,340	22,728
1901/1902	2,496,494	1,746,121	750,373	1934/1935	4,626,220	3,883,304	742,916
1902/1903	2,592,779	1,826,355	766,424	1935/1936	4,289,611	4,074,358	215,253
1903/1904	2,690,571	1,792,426	898,145	1936/1937	4,635,742	4,049,121	586,621
1904/1905	2,828,171	2,084,480	743,691	1937/1938	4,857,841	4,095,373	762,468
1905/1906	3,036,892	2,316,731	720,161	1938/1939	4,886,944	3,998,784	888,160

(*continued*)

Table 3.10 (continued)

Year	*Revenue*	*Expenditure*	*Net income*	*Year*	*Revenue*	*Expenditure*	*Net income*
1906/1907	3,450,733	2,586,566	864,167	1939/1940	4,292,000	3,857,000	435,000
1907/1908	3,858,026	2,823,433	1,034,593	1940/1941	4,634,567	3,735,903	898,664
1908/1909	3,886,996	2,985,358	901,638	1941/1942	4,891,402	3,655,578	1,235,824
1909/1910	4,184,203	3,384,927	799,276	1942/1943	6,977,674	4,966,868	2,010,806
1910/1911	3,791,940	3,218,552	573,388	1943/1944	11,299,940	7,986,901	3,313,039
1911/1912	4,237,795	3,446,441	791,354	1944/1945	14,074,827	8,148,016	5,926,811
1912/1913	4,118,771	3,042,157	1,076,614	1945/1946	14,012,541	7,639,884	6,372,657
1914/1915	4,048,842	2,840,942	1,207,900	1946/1947	15,867,746	7,971,742	7,896,004

Source: C.A.R. Bhadran, 'Forests of the Madras State' in *Forest Centenary of Forest Administration in Madras State*, Madras: Forest Department, 1959, pp. x–xi and *Madras Administration Report* (for various years)

deterioration in the tribals' livelihood patterns could be observed from the following variables: the increasing number of non-tribal settlements, tribal land alienation, land rent, grazing fees, forest-related offences, and a decline in occupational status and the landholding pattern.

Population Growth in the Tribal Areas

Population growth steadily rose in the different hills of the Madras Presidency from the late nineteenth century to the first half of the twentieth century, the post-Forest Act colonial period. Let us see the demographic trend in five hills (Attur-Kalrayan, Pachmalai, Attur-Kolli, Namakkal-Kolli, Shervaroy hills) in the north-western parts of the Madras Presidency during the period. In these five hills, the population increased from 36,566 in 1881 to 50,847 in 1951.[53] The data clearly indicates that the population growth was about 39% during these seven decades. In the major hills, the population growth was comparatively higher than in the smaller hills. For instance, in the Shervaroy hills, the growth rate was about 88% and in the Kalrayan hills it was about 58%; in the Kolli hills it was around 7%. However, in the first two decades of the twentieth century, the population in the hills remained constant owing to a severe famine, with people also dying because of fever, malaria, cholera and smallpox.[54] Indeed, during the first three decades of the twentieth century, the tribal population decreased owing to these natural calamities. As such, the population growth in the hills in the subsequent decades could be attributed to the large-scale non-tribal settlements.

Non-Tribal Settlement

The presence of non-tribals in the hills was insignificant until the early nineteenth century. From the first quarter of that century the British started settling in the hills, particularly where plantations were established. However, until the 1871 Census there was no separate data for non-tribals in the hills. To understand the trend of non-tribal settlements in the tribal areas of the Madras Presidency, Salem district was selected because it had the largest number of tribals. According to the 1871 Census, about 4026 persons or about 12% of the total population in the five major hills were found to be non-tribals.[55] Of them, 2649 persons or about 65.80% were *Pariahs* (a Scheduled Caste under the Indian constitution), who were bonded labourers in the plains.[56] Only seven persons or about 0.17% were

Chetties; 26 persons or about 0.65% were Muslims and the remaining 1343 persons or about 34% belonged to other castes. The non-tribals largely settled in the Shervaroy hills, Namakkal-Kolli hills, Attur-Kolli hills and Kalrayan hills, and their percentage to the total population in these areas was respectively 21.30%, 10.37%, 9.16% and 4.5%. These people migrated to the hills to take advantage of the low land prices. In the large hills, the problem of land alienation raised its head during the second half of the nineteenth century. An exception was the Shervaroy hills, where this problem emerged only in the second quarter of the century.[57] Therefore, it should be noted that land alienation existed only in the larger hills; no such problem was experienced in the smaller hills.

The data pertaining to non-tribals in the hills is not available in subsequent censuses until 1961. However, the density of population increased in various hill regions during the early twentieth century, indicating a spurt in non-tribal settlements. For example, in the Shervaroy hills the population density per square mile increased from 498 in 1911 to 618 in 1931.[58] According to the 1961 Census, the total number of non-tribals in various hills had increased to 11,148, or about 22% of the total population of the hills.[59] In other words, the proportion of non-tribals increased from 12% in 1871 to 22% in 1961, with a growth rate of about 177%. In short, the number of non-tribals, particularly Scheduled Castes population, in the various hills increased in the Madras Presidency during the post-Forest Act colonial period, which could have been played a significant role in the alienation of tribal lands.

Land Alienation

The influx of non-tribals in various hill areas led to the alienation of tribal land. In the early twentieth century, about 6.69% of the land was occupied by the non-tribals, particularly by planters and *sowcars.*[60] In the Shervaroy hills, about 1417.08 acres or 11.7% of lands were alienated; of which planters occupied 918.11 acres or 64.79%. In Attur-Kolli hills, about 584.66 acres or 6.13% of the total occupied land was alienated by the non-tribals, mainly for agricultural purposes. In the Namakkal-Kolli hills, about 289.26 acres or about 2.29% of the total land was taken over by the non-tribals. Of this, 95.97% was alienated by the *sowcars.* In Pachamalai, about 300 acres of land had changed hands in 1925 owing to failures to repay loans.[61] The tribal lands in the hill areas were transferred not only to the non-tribals but also to some extent among the tribals themselves.[62]

Economic Decline

Until the advent of the colonial administration, and even later, the tribal economy of the Madras Presidency was characterised by traditional methods of cultivation and collection of forest produce.[63] Indeed, tribals were classified with the occupational category of cultivator in 1836.[64] Owing to the intervention of the British settlers and other non-tribals and the introduction of coffee plantations, their occupational status declined from the early nineteenth century.[65] After the Madras Forest Act 1882, the economic conditions of the tribals deteriorated owing to the introduction of the reserve forest, the intervention of non-tribals and the alienation of their land. For instance, in various hills of the Madras Presidency little more than about 7% (2503) of the total population (34,366) was classified as labourers in 1871. Invariably, the proportion of labourers was higher in the major hills.[66] In the Shervaroy hills, about 8% of the total population was classified thus, and this necessarily includes the plantation sector. It is evident that in the Shervaroy hills that displacement from their lands had turned a large proportion of tribals into agricultural and plantation workers.[67]

After the introduction of the Forest Act, the proportion of non-tribals increased. According to the 1951 Census, about 20% (10,044) of the total population (50,325) was classified as labourers and their dependents,[68] and nearly half of them were female.[69] The evidence clearly suggests that the tribal-used forest resources were brought into the reserve forest, while a large number of non-tribals were also settled in the hill areas, leading to a decline in their economic status. The proportion of labourers to the total population progressively increased during the post-Forest Act period.

Extent, Holding Pattern and Assessment of Land

The economic decline of the tribals can be ascertained from the trends pertaining to landholding pattern, extent of land and collection of land revenue during the post-Forest Act period. Subsistence livelihood was the basic characteristic feature of the tribal economy, with the tribals occupying small landholdings only with occupancy rights. In fact, prior to the survey and settlement (1904), tribals paid land revenue based on the number of implements used for cultivation and not on the extent of the land they cultivated.[70] The size of landholdings can be seen from the amount of rent paid in different hills in the Madras Presidency, in which more than 90% of the tribal households had small landholdings.[71] The forests were brought into the reserve forest under the Madras Forest Act during the last quarter of the nineteenth century, and then the survey and settlement were introduced in the different hill regions during the early twentieth century.

Prior to the survey and settlement, it was estimated that about 22,266 acres of land was cultivated and this increased to 37,261 acres under survey and settlement, which was about a 67.34% increase. During the resurvey and settlement in 1934, the amount of land had increased to 38,289 acres or about 79.59% more than prior to survey and settlement, or about 7.32% more than at first survey and settlement (Table 3.11). This clearly indicates that more and more land was being brought under cultivation during the early twentieth century. Though the cultivated area had increased, the number of holdings also increased during this period. According to the survey and settlement (1904), the number of holdings was about 6057 of which nearly 90% of the owners paid less than Rs.10, a clear proof that a vast majority of the tribals had only small and marginal holdings. Subsequently, there was an increase, and in 1929 the total number of landholdings was about 7919, an increase of 1862 or 30.74% when compared with 1904 (Table 3.11).

During the early twentieth century, not only the extent of land and the number of holdings increased but also the land revenue, which registered a sharp rise during the period. Prior to the survey and settlement, the land revenue realised was about Rs.21,626 but this increased to Rs.24,317 or by about 12.44% during the survey and settlement. During the resurvey and settlement in 1932, it had further increased to Rs.30,260, about 40% more than the revenue settlement and about 24.44% higher than the first survey and settlement (Table 3.11). Although there was an expansion of the area under cultivation, an increase in landholdings as well as rent resulted in the marginalisation of the tribals.

Cropping Pattern

Except in the Shervaroy hills, where coffee plantations were introduced during the early nineteenth century, the cropping pattern had not changed in the other hills of the Madras Presidency and the tribals continued to cultivate foodgrains. The main crops cultivated in the hills of Salem and Baramahal regions until 1905 settlement were *ragy* or *kevaroo*, *thinay*, *varagu*, *cholam*, *kambu*, *panivaragu* and *dry-paddy*; pulses—*mochaikottay*, *thovarai*, *tatapyru*, *kodallay* and *avarai*, and condiments—mustards, chillies, onion and garlic. In 1904, about 86.62% of the

Table 3.11 Extent, holding pattern and assessment of land in the Madras Presidency, 1900–1932

Name of the hills	*Extent (in acres)*				*No. of holdings*			*Assessment (in rupees)*			
	Prior to survey and settlement	*1904*	*1912*	*1932*	*1904*	*1914*	*1929*	*Prior to survey and settlement*	*1905*	*1912*	*1932*
Pachamalai	1030	3022	3647	3681	192	343	381	944	1419	1403	1417
Kolli hills	11,550	22,159	22,259	22,692	4185	4398	5380	11,000	11,794	16,440	16,615
Shervaroy hills	9686	12,080	12,383	13,614	1680	1763	2158	9682	11,104	11,188	12,228
Total	22,266	37,261	38,289	39,987	6057	6504	7919	21,626	24,317	29,031	30,260

Sources: G.O. No. 605, Revenue, 30 June 1905, TNSA; G.O. No. 1793, Revenue, 17 August 1936, pp. 66, TNSA

total cropped area in different hills was under foodgrains; 7.09% under pulses; 3.6% constituting oil-seeds and only 2.7% other crops, coffee, tobacco, fruits and vegetables.[72] There was not much of a change in the cropping pattern in the different hills between the settlement periods.[73] In fact, the area under foodgrains cultivation had increased in most of the hills. The area under other crops had also increased marginally in the various hills except Shervaroys, where coffee was cultivated to some extent mainly by the British settlers and others from the plains. Without any remarkable change in their income sources, the inhabitants' economic status had deteriorated in different hills of the Madras Presidency during the post-Forest Act period.

Trends of Forest-Related Offences

The Madras Forest Act 1882 assumes significance in that it brought the tribals under the modern legal system. In other words, it sounded the death-knell for the traditional customary practices of the tribal communities and other forest users, which were notified as offences. Corresponding to the increase in the extent of reserve forests, there was a rise in forest-related offences. For instance, only 2493 offences were reported in 1886/1887, rising to 13,218 in 1896/1897. This continued to move upwards in the subsequent period (Table 3.12). It indicates that more and more tribals were brought under the purview of law and were subjected to the attendant harassment.

Impact of on Grazing

Consequent to the closure of vast tracts of forests to grazing and a restriction on the number of animals, the post-Forest Act period witnessed a number of animals being impounded and a huge amount was realised as penalties through the court for forest-related offences during this period. For instance, in 1901/1902, about 4.66 lakh animals were impounded.

In addition to this, a huge amount was realised as compounding fees through the court for forest-related offences during this period (Table 3.13). In the ensuring years this decreased, subsequent to a relaxation in the restrictions and the transfer of a sizeable amount of forest to the forest *panchayat*. Not only were the tribals' livelihood sources taken over and their cattle resources curtailed, but they were also brought under the legal process which involved penalties as well as court procedures.

Table 3.12 Trends in the number of offences, 1886/1887–1939–1940

Year	*No. of offences*	*Year*	*No. of offences*	*Year*	*No. of offences*
1886/1887	2493	1904/1905	21,930	1922/1923	30,878
1887/1888	3358	1905/1906	23,021	1923/1924	30,205
1888/1889	3934	1906/1907	23,141	1924/1925	32,420
1889/1890	4034	1907/1908	24,453	1925/1926	29,940
1890/1891	5577	1908/1909	25,470	1926/1927	25,321
1891/1892	N.A	1909/1910	25,277	1927/1928	25,976
1892/1893	N.A	1910/1911	26,498	1928/1929	24,714
1893/1894	10,007	1911/1912	29,222	1929/1930	23,345
1894/1895	9883	1912/1913	29,892	1930/1931	24,798
1895/1896	9610	1913/1914	31,777	1931/1932	23,876
1896/1897	13,218	1914/1915	32,541	1932/1933	29,476
1897/1898	16,769	1915/1916	30,885	1933/1934	30,616
1898/1899	20,495	1916/1917	28,087	1934/1935	28,242
1899/1900	22,810	1917/1918	27,091	1935/1936	28,936
1900/1901	22,130	1918/1919	26,160	1936/1937	27,017
1901/1902	23,750	1919/1920	27,537	1937/1938	26,738
1902/1903	23,125	1920/1921	23,947	1938/1939	28,890
1903/1904	21,883	1921/1922	29,040	1939/1940	31,221

Sources: *Administrative Report of the Madras Presidency* and *Administration Report of the Forest Department of the Madras Presidency* (various years)

N.A not available

Conclusion

Until recently, the focus of the debate on the environmental history of India was primarily centred on the pre-Forest Act period, while the post-Forest Act era, especially under colonial rule, was largely neglected. The Madras Forest Act 1882, while failing to provide a required impetus to conservation, had only crippled the tribal economy. It was in spite of the law that the colonial state carried out a programme of commercialisation. Even the initiatives directed towards environmental protection were guided by the need to sustain resources for future exploitation. In order to ensure sustainability, the Forest Act imposed a whole range of restrictions on the tribals and other forest users from the late nineteenth century.

Even while curtailing the rights of the tribals, which had been enjoyed from time immemorial, the government allowed commercial exploitation to take strong root. Colonial propaganda cleverly shifted the blame to the tribals and other forest dwellers for the denudation and degradation of the

Table 3.13 Number of animals impounded and compounding fees collected in the Madras Presidency, 1901/1902–1946/1947

Year	*No. of animals impounded*	*Compounding fees (in rupees)*	*Year*	*No. of animals impounded*	*Compounding fees (in rupees)*	*Year*	*No. of animals impounded*	*Compounding fees (in rupees)*
1901/1902	465,542	97,428	1917/1918	148,806	173,000	1933/1934	57,459	168,209
1902/1903	370,388	84,556	1918/1919	N.A	157,000	1934/1935	54,776	144,884
1903/1904	292,196	90,784	1919/1920	N.A	205,000	1935/1936	54,068	140,530
1904/1905	228,322	102,444	1920/1921	N.A	179,000	1936/1937	53,152	142,618
1905/1906	230,796	116,858	1921/1922	N.A	229,633	1937/1938	43,887	134,941
1906/1907	204,678	139,028	1922/1923	N.A	272,080	1938/1939	48,010	111,742
1907/1908	191,039	135,146	1923/1924	N.A	281,310	1939/1940	N.A	132,000
1908/1909	N.A	169,686	1924/1925	N.A	265,310	1940/1941	N.A	N.A
1909/1910	209,930	153,715	1925/1926	144,882	268,404	1941/1942	N.A	N.A
1910/1911	202,093	169,686	1926/1927	83,798	204,339	1942/1943	N.A	N.A
1911/1912	200,161	169,461	1927/1928	77,076	184,468	1943/1944	N.A	N.A
1912/1913	196,108	194,940	1928/1929	59,487	174,381	1944/1945	N.A	N.A
1913/1914	190,714	214,180	1929/1930	60,254	199,238	1945/1946	35,165	218,000
1914/1915	208,393	208,959	1930/1931	59,082	193,077	1946/1947	40,189	236,000
1915/1916	175,903	N.A	1931/1932	76,370	166,000			
1916/1917	222,367	N.A	1932/1933	N.A	168,629			

Source: Administrative Report of the Madras Presidency and Administration Report of the Forest Department of the Madras Presidency (various years)

N.A not available

green cover. This was nothing but a ploy to cover up the colonial project to plunder natural wealth without any accountability. The massive exploitation of forest resources continued unhindered even after the 1882 Act came into force, and this had reached its pinnacle during the Second World War.

Development of infrastructure such as roads, though very limited, in the hills during the colonial period facilitated the extraction of forest products. There was nothing in place to act as a catalyst to help promote the tribal economy, for modern infrastructure facilities remained inaccessible to indigenous communities and inhabitants. The colonial government's initiatives towards conservation were not designed to preserve the environment and natural resources but rather to ensure the supply of forest products, not only in the present but in the future as well. Hence, concerns related to the environment and forest ecology were left unaddressed or were ignored, and the situation was allowed to deteriorate. Corresponding to this was the disintegration of the traditional tribal system and the decline in tribal living conditions. Policies pursued during the post-Forest Act British rule have neither protected the environment nor promoted the tribals' livelihood, their options shrinking all the time. Rather, they have paved the way for massive non-tribal settlements and attendant land alienation, the commercialisation of the forests and the striking down of the traditional bond between the forests and the tribals.

Notes

1. About 73 square miles in the Anamalai hills were given up to planters (Stebbing Vol. IV; p. 175).
2. Dietrich Brandis, *Suggestions Regarding Forest Administration in the Madras Presidency*, Madras: Government Press, 1983, p. 4.
3. Velayutham Saravanan, 'Tribal Land Alienation in Madras Presidency during the Colonial Period: 1792–1947'. *Review of Development and Change*, 6(1), 2000, p. 72; Velayutham Saravanan, 'Colonial Agrarian Policies in the Tribal Areas of Madras Presidency: 1872–1947'. *South Asia Research*, 26(1), 2006, p. 72.
4. Government of Madras, *Administrative Report of the Madras Presidency* 1933–34, Madras: Government Press, p. 95.
5. E.P. Stebbing, *The Forests of India*, Vol. IV, London: Oxford University Press, 1962, p. 174.
6. Ibid., pp. 174–175.
7. Ibid., p. 175.

8. Ibid., p. 176.
9. Ibid., p. 175.
10. Velayutham Saravanan, 'Commercialisation of Forest, Environmental Negligence and Alienation of Tribal Rights in Madras Presidency: 1792–1882'. *Indian Economic and Social History Review*, 35(2), 1998, pp. 125–146; Velayutham Saravanan, 'Colonialism and Coffee Plantations: Decline of Environment and Tribals in Madras Presidency during the 19th Century'. *Indian Economic and Social History Review*, 41(4), 2004, pp. 465–488.
11. Velayutham Saravanan, 'Commercial Crops, Alienation of Common Property Resources and Change in Tribal Economy in the Shervaroy Hills of Madras Presidency During the Colonial Period', *Review of Development and Change*, 4(2), 1999, pp. 298–317; Velayutham Saravanan, 'Economic Exploitation of Forest Resources in South India during Pre-Forest Act Colonial Era, 1793–1882,' *International Forestry Review: The International Journal of Forest Science and Policy*, 10(1), 2008, pp. 65–73.
12. See Madras Administration Report (various years).
13. Stebbing, *The Forests of India*, p. 177.
14. See Administration Report of the Madras Presidency and Administration Report of the Forest Department (various years).
15. W.H.W. Adolphus, *Consolidated Working Plan for the Central Salem Division*, Madras, 1932, p. 19.
16. Government Order (hereafter G. O.), No.1753, Revenue, 17 August 1936, p. 10 (Tamil Nadu State Archives, Chennai, hereafter TNSA).
17. It was a lowest money measurement: 12 *pies* equal to one *anna*, 16 *annas* or 172 *pies* equal to Company rupees.
18. G. O. No. 1139 [Misc], Development, 27 January, 1921, TNSA.
19. G. O. No. 1793, Revenue, 17 August, 1936, p. 9, TNSA.
20. Ibid., p. 10.
21. Ibid.
22. Ibid.
23. Ibid.
24. Ibid.
25. Saravanan, 'Commercialisation of Forest, Environmental Negligence, pp. 125–146; Saravanan, 'Economic Exploitation of Forest Resources in South India'.
26. Saravanan, 'Commercialisation of Forest, Environmental Negligence and Alienation of Tribal rights', pp. 125–146; Saravanan, 'Colonialism and Coffee Plantations', pp. 465–488; Velayutham Saravanan, 'Colonial Commercial Forest Policy and the Tribal Private Forests in Madras Presidency: 1792–1881'. *Indian Economic and Social History Review*, 40(4), 2003, pp. 403–423.

27. *Report on the Administration of the Forest Department*, 1923–1924, p. 55.
28. Saravanan, 'Commercialisation of Forest, Environmental Negligence, pp. 125–146; Saravanan, 'Economic Exploitation of Forest Resources in South India'.
29. Ramachandra Guha and Madhav Gadgil, 'State Forestry and Social Conflict in British India'. *Past and Present*, 122, 1989, pp. 120–123.
30. Morris David Morris and Clyde B. Dudley, 'Selected Railway Statistics for the Indian Subcontinent (India, Pakistan and Bangladesh, 1853–1946–47' *Artha Vijnana*, 17(3), 1975, pp. 194–196.
31. Ibid., pp. 243–245.
32. Saravanan, 'Economic Exploitation of Forest Resources in South India'.
33. Statistical Abstract Relating to British India (various Years).
34. Brandis, '*Suggestions Regarding Forest Administration'*, p. 32.
35. Ibid., p. 40.
36. Saravanan, 'Commercialisation of Forest, Environmental Negligence, pp. 125–146; 'Economic Exploitation of Forest Resources in South India'.
37. Saravanan, 'Commercialisation of Forest, Environmental Negligence, pp. 125–146.
38. Madras Administration Reports (for various years).
39. Velayutham Saravanan, 'Environmental History of Tamil Nadu State, Law and Decline of Forest and Tribals, 1950–2000'. *Modern Asian Studies*, 41(4), 2007, pp. 723–767.
40. Saravanan, 'Colonialism and Coffee Plantations', pp. 465–488.
41. C.D. Maclean, *Manual of the Administration of the Madras Presidency*, 3 Vols. Madras: Government Press, 1885, p. 290.
42. Dharma Kumar, *Land and Caste in South India: agricultural labour in Madras Presidency in the nineteenth century*, Cambridge: Cambridge University Press, 1965, p. 126.
43. Maclean, *Manual of the Administration of the Madras Presidency*, p. 404.
44. G. O. No. 1793, Revenue, 17 August 1936, p. 49, TNSA.
45. Government of Madras, *Season and Crop Report of the Madras Presidency 1912–13*, Madras: Government Press.
46. Maclean, *Manual of the Administration of the Madras Presidency*, p. 336.
47. Saravanan, 'Commercialisation of Forest, Environmental Negligence, pp. 125–146; Saravanan, 'Economic Exploitation of Forest Resources in South India'.
48. Government of Madras, *Administration Report of the Forest Department, 1927–28,* Madras: Government Press, p. 7.
49. Stebbing, *The Forests of India*, p. 177.
50. Ibid., p. 186.
51. Ibid., p. 187.

52. Saravanan, 'Commercialisation of Forest, Environmental Negligence, pp. 125–146; Saravanan, 'Economic Exploitation of Forest Resources in South India'; Saravanan, 'Tribal Land Alienation in Madras Presidency'; Saravanan, 'Colonialism and Coffee Plantations: Decline of Environment and Tribals'; Velayutham Saravanan, 'Agrarian policies in the tribal areas of Madras Presidency during the pre–survey and settlement period, 1792–1872, *Indian Journal of Agricultural Economics*, 65(2), 2010, pp. 261–276; Velayutham Saravanan, 'Thamizhaga Malaival Makkalin Nila Urimai' ['Land Rights of Tribals in Tamil Nadu'. Special Issue]. *Uzhavan Urimai*: 1999, pp. 53–57 (in Tamil).
53. Census Statement of Population of 1871, Salem District and Census of India 1951, *Census Handbook, Salem District*.
54. G. O. No. 2332 [Mis], Revenue, 24 October 1925; G. O. No. 1793, Revenue, 17 August, 1936, pp. 17–45, TNSA.
55. Census Statement of Population of 1871, Salem District.
56. 'In times prior to colonial rule the whole of the Pariah community, without exception, were the slaves of the superior castes' (Census of India 1871. *Census of the Madras Presidency 1871 A Report*, Vol. I, Madras, 1874, p. 169.
57. Saravanan, 'Tribal Land Alienation in Madras Presidency'; Saravanan, 'Commercial Crops, Alienation of Common Property Resources'.
58. G. O. No. 1793, Revenue, 17 August 1936, p. 91, TNSA.
59. Census of India 1961, Vol. IX, Madras Part XXI. District Census Hand Book, Salem, Vol. II, 1965.
60. G. O. No. 605, Revenue, 30 June 1905, pp. 15, 19, 27, and 33, TNSA; G. O. No. 1130, Revenue, 18 May 1916, TNSA.
61. G. O. No. 2332 [Misc], Revenue, 24 October 1925, p. 3, TNSA.
62. Institute of Techno–Economic Studies, *Indebtedness of Scheduled Tribes in Tamil Nadu,* Madras: Institute of Techno–Economic Studies, 1978.
63. Saravanan, 'Commercial Crops, Alienation of Common Property Resources', p. 304.
64. BOR, Vol. 1537, 10 November. 1836, pp. 18133–18136, TNSA.
65. Saravanan, 'Colonialism and Coffee Plantations: Decline of Environment and Tribals'.
66. Census Statement of Population of 1871, Salem District.
67. Ibid.
68. In the 1951 Census, the cultivators of land wholly or mainly unowned and their dependants, cultivating labourers and their dependants are brought under the labourers category.
69. Census of India 1951, Census Handbook, Salem District.

70. Velayutham Saravanan, 'Agrarian policies in the tribal areas of Madras Presidency during the pre-survey and settlement period, 1792–1872, *Indian Journal of Agricultural Economics*, 65(2), 2010, pp. 261–276.
71. G. O. No. 605, Revenue, 30 June 1905, TNSA.
72. G. O. No. 1793, Revenue, 17 August 1936, TNSA.
73. Ibid.

Check for updates

CHAPTER 4

Forests and Tribals, 1950–2000

Environment and sustainable development have been accorded great emphasis since the last quarter of the twentieth century. In India, the environmental protection is enshrined in the Constitution of India (42nd Amendment) under the Directive Principles of State Policy of 1977. According to Article 48A, 'State shall endeavour to protect and improve the environment and to safeguard the forests and wildlife in the country.' Article 51A(g) enjoins upon the citizens 'to protect and improve the natural environment including forests, lakes and river and wildlife and to have compassion for the living creatures'.

Forestry is one of the core components of the environment, which constitutes biodiversity and tribal life. In the process of development, deforestation and denudation have become a common phenomenon, affecting sustainability. For instance, about 30% of the total forest reserve (2,144,065 hectares) in Tamil Nadu, a part of the erstwhile Madras Presidency in the Indian subcontinent, was classified as degraded forest by 1993.[1] In such a grave situation, an attempt to study the effects of post-colonial state policy on the environment and tribals in Tamil Nadu will help to mould a sustainable environment for development.

Until independence, in fact until the last quarter of the twentieth century, scholarly works were found wanting regarding issues concerned with forests and forest-dwellers as well as forest-users. With the emergence of

This chapter was originally published in *Modern Asian Studies*, Vol. 41(4), 2007.

© The Author(s) 2018

V. Saravanan, *Environmental History and Tribals in Modern India*,
https://doi.org/10.1007/978-981-10-8052-4_4

environmental movements in many parts of the world and in India in the 1970s, scholarly debates were initiated during the last quarter of the twentieth century, particularly in the 1980s. However, the environmental history debate concentrated mainly on the colonial period, particularly the nineteenth century.[2] Recent studies on the environmental history of the Madras Presidency have brought out the dynamic link between the forests and the tribals from a historical perspective.[3] These studies have found that attempts to encroach upon tribal-controlled regions faced stiff resistance.[4] The tribals launched several agitations and revolts during the eighteenth century to ward off external control.[5] Further, it is evident that until the introduction of the Madras Forest Act 1882, the colonial government had neither considered issues concerning environmental degradation, nor desisted from trampling upon the tribals' rights over the forests. It is appalling to note that no welfare measure was initiated during the period.[6] In fact, the colonial government accorded priority and emphasis only to the commercialisation of forests, which gradually led to disruption of the tribal life and settlements.[7] When coffee plantations were introduced in the hill areas, prior to the implementation of the Madras Forest Act, the British planters alone were encouraged to engage in the venture.[8] Even the Act of 1882, which sought to extend the area under reserve forests as a strong measure for ecological and environmental conservation, placed several curbs on the tribals' customary rights. Subsequently, in the early twentieth century, lands traditionally owned by the tribals with occupancy rights were surveyed and legal rights were given only to those lands that were not in the reserve forests area.[9] The Forest Department further restricted the tribals' rights through the subsequent Forest Acts. Further, state institutions such as the Forest, Revenue and Police Department authorities who were responsible for executing these Acts digressed from their due responsibility and violated human rights in different tribal pockets of Tamil Nadu.[10] In addition, the influx of a large number of non-tribals who occupied the hills resulted in the flight of tribals from their lands.[11]

Studies pertaining to the post-colonial period invariably focus either on the impact of the welfare programmes or on the socio-cultural aspects of the tribals in the hilly regions.[12] However, a few have attempted to look at environmental issues on the micro-level.[13] The impact of post-colonial projects on the environment has been almost ignored. Given this scenario, the present chapter attempts to analyse the environmental history of Tamil Nadu during the post-independence period.

In Tamil Nadu, contemporary environmental problems are mainly associated with three dominant actors: state, tribals and non-tribals. First, the state, while being committed to conservation, has enacted several Forest Acts and yet continued to encroach on forests for various developmental activities. The common thread running through all the forest policies, formulated over the years, has been the encouragement offered to commercialisation of forest resources. Second, the cultural and economic life of the tribals has since antiquity remained forest-oriented, with their world view being moulded by the hills. Over the period under study, several restrictions have been enforced through a number of Forest Acts. In spite of preventive measures, forest wealth has continued to show a sharp decline. Yet it is still to be proved whether the tribals were responsible for deforestation in the contemporary period. However, that tribal development initiatives have changed their conventional forest-dependent livelihood into a resourceful mode of life cannot be overlooked. Third, the role of non-tribals, more importantly the politicians and contractors who have colluded to exploit forest resources through law enforcing machinery, needs an elaborate study to ascertain whether lacunae have arisen from their policies.

The main objective of this chapter is to study the forest policies of successive governments and their impact on the environment and the tribal economy in Tamil Nadu during the second half of the twentieth century. The study therefore has to cover certain important research questions to ascertain how the post-colonial policy has altered the landscape, forests and tribals, and had irreversible consequences on the environment. To what extent have the policy implementation and the major developmental projects created adverse impacts? Has environmental protection been accorded due importance in the developmental agenda or has the state merely been interested in garnering more revenue through commercialisation? What are the other factors responsible for the depletion of forest resources? In placing emphasis on commercialisation, has the state considered providing alternatives for the tribals? These questions might be helpful in understanding state policy in regard to the tribals, environment and changes in land use patterns (if any) during the study period.

The focus is on the second half of the twentieth century for three important reasons. First, despite the extension of reserve forests, forest cover has shrunk. Second, while restrictions on tribals' rights over forests have been strictly enforced, forests have been taken over by the state under one pretext or the other. Third, infrastructure development and the attendant consequences on the tribal economy as well as catastrophic environmental damage are still prominent.

Trends of Forest Areas in Tamil Nadu

Trends of Land Utilisation

Land utilisation trends in Tamil Nadu have changed during the post-independence period. Barren and uncultivable land, cultivable waste and permanent pastures have constantly witnessed a sharp decline during the last five decades. This is primarily due to extensive afforestation measures such as farm forestry, social forestry and other programmes, all aimed at developing these areas to cater to the consumable needs of the local people. Consequently, the common property resources mentioned above have started vanishing at an alarming rate during the first three decades of the post-independence period: it has been estimated that about 50% have disappeared in Tamil Nadu in this period.[14] The sharp decline has been primarily attributed to gradual but steady extension of private field borders into the common property resources both individually and collectively, government distribution of common property lands to landless people under land reform programmes and the encroachment of various government departments upon the forest tracts for development activities. At the same time, there has been a quantum leap in turning cultivable land over to non-agricultural purposes during this period owing to the ever-expanding process of urbanisation. Land occupied by buildings, roads, canals and industrial plants besides that put under non-agricultural use is brought under this category. Ironically, the area categorised as fallow lands has gone up, particularly during the post-Green Revolution period. Since the early 1990s, the technology-driven Green Revolution has not only led to a scarcity of water resources but also resulted in an increase in fallow lands. However, the area under forest cover has risen together with the other types of uncultivable lands that are categorised with forests. Throughout this period, the common property resources with a specific use have been slowly eliminated to a large extent throughout the state.

Expansion of Agriculture

Immediately after independence, the government encouraged the extension of cultivation under the Grow More Food campaigns, which resulted in the large-scale clearing of forest in Tamil Nadu.[15] The large-scale expansion of agriculture continued with greater institutional support during the first three decades of the late twentieth century. This was further facilitated by the introduction of modern technology during the Green Revolution that changed the face of agricultural operations. Tractors, harvesting machines, pump-sets, oil-engines and institutional support, including credit facilities, have created a conducive climate for the rapid expansion of agriculture. Consequently, the area under cultivation increased many times over during the 1960s and early 1970s: the net cultivated area was 5,898,000 hectares in 1960/1961 and increased to 6,269,000 hectares in 1970/1971 (Table 4.1). In subsequent years, this trend reversed, with the net cultivated area registering a decline in correspondence to the rise in fallow lands. The undeniable fact is that the Green Revolution technologies and institutional support that led to the depletion of water tables has resulted in an increase in fallow lands. In short, while more and more common property resources were initially brought under cultivation, this momentum could not be sustained owing to a depletion of resources that resulted in the decline of the net irrigated area during the post-Green Revolution period.

Trends of Forest Area

Since independence, the area under forest cover has increased from 1,866,000 hectares in 1960/1961 to 2,134,000 hectares in 1999/2000 or 14.13% to 16.55% of the total geographical area (Table 4.2). The villages' common property resources such as barren and cultivable waste and pastoral lands has been reclassified as forest lands after several afforestation programmes have been taken up. The increase in area under forest cover witnesses fluctuations almost every year which cannot be brushed aside. In the first five decades, the rise and fall were nearly equal. While there was a rise almost for the first half of the period, the later stages show a decline. The latter is an indication that forest land was diverted for development projects and rehabilitation purposes. Even after the enactment of the Conservation of Forest Act in 1980, a sizeable amount of forest has been diverted to purposes other than forest use. On the whole, about 321,000 hectares have been brought under forest cover, while about 202,000 ha were diverted between 1951/1952 and 1999/2000 (Table 4.2). Some of the *Jagir* (Ex-zamin) forests in the erstwhile South Arcot, Madurai, Virudunagar, Dindigul, Nilgiris and Tiruchirappalli districts were brought into the reserve forests. Further, 22,496 hectares in 1987/1988, 29,483 hectares in 1988/1989 and 6454 hectares in 1989/1990 (up to February 1990) were annexed to the reserve forests.[16] Hence, it becomes evident that despite the apparent attempts of the state to expand forest cover, a very large amount of forest land was diverted for different developmental activities.

Table 4.1 Trends of land utilisation in Tamil Nadu, 1960–2000 (area in Thousand hectares)

Classification	*Area in hectares*						*% to total area*				
	1960/ 1961	*1970/ 1971*	*1980/ 1981*	*1990/ 1991*	*1999/ 1900*	*Change 1999/1900– 1960/1961*	*1960/ 1961*	*1970/ 1971*	*1980/ 1981*	*1990/ 1991*	*1999/ 1900*
Forests	1866	2013	2022	2155	2134	268	14.34	15.36	15.55	16.55	16.43
Barren & uncultivable land	945	832	577	509	476	–469	7.26	6.35	4.44	3.91	3.66
Land put to non-agrl. use	1295	1489	1747	1820	1978	683	9.95	11.36	13.44	13.98	15.22
Cultivable waste	706	507	343	290	349	–357	5.42	3.87	2.64	2.23	2.69
Permanent pasture	363	231	160	124	123	–240	2.79	1.76	1.23	0.95	0.95
Misc. tree crops	246	226	213	234	243	–3	1.89	1.72	1.64	1.80	1.87
Current fallows	974	964	2121	1264	1085	111	7.48	7.36	16.31	9.71	8.35
Others fallows	623	573	459	1044	1140	517	4.79	4.37	3.53	8.02	8.77
Net area cultivated	5997	6269	5360	5578	5464	–533	46.08	47.84	41.22	42.85	42.06
Total geographical area	13,015	13,104	13,002	13,018	12,992	–23	100.00	100.00	100.00	100.00	100.00

Source: Tamil Nadu: an economic appraisal (various years)

Table 4.2 Trends of forest area in Tamil Nadu, 1951/1952–1999/2000 (area in Thousand hectares)

Year	*Area under forest (in ha)*	*Increase/ decrease*	*Year*	*Area under forest (in ha)*	*Increase/ decrease*
1951–1952	1813	–	1976–1977	2000	+28
1952–1953	1831	+18	1977–1978	2023	+23
1953–1954	1835	+4	1978–1979	2031	+8
1954–1955	1817	−18	1979–1980	2022	−9
1955–1956	1816	−1	1980–1981	2023	+1
1956–1957	1788	−28	1981–1982	2024	+1
1957–1958	1777	−11	1982–1983	2030	+6
1958–1959	1775	−2	1983–1984	2030	0
1959–1960	1878	+103	1984–1985	2050	+20
1960–1961	1866	−12	1985–1986	2066	+16
1961–1962	1883	+17	1986–1987	2091	+25
1962–1963	1868	−15	1987–1988	2122	+31
1963–1964	1885	+17	1988–1989	2164	+42
1964–1965	1866	−19	1989–1990	2166	+2
1965–1966	1907	+41	1990–1991	2155	−11
1966–1967	1916	+9	1991–1992	2147	−8
1967–1968	1924	+8	1992–1993	2151	+4
1968–1969	1925	+1	1993–1994	2144	−7
1969–1970	2016	+91	1994–1995	2144	0
1970–1971	2013	−3	1995–1996	2143	−1
1971–1972	2007	−6	1996–1997	2141	−2
1972–1973	2008	+1	1997–1998	2140	−1
1973–1974	2008	0	1998–1999	2140	0
1974–1975	1966	−42	1999–2000	2134	−6
1975–1976	1972	+6			

Source: Tamil Nadu: an economic appraisal (various years)

Trends of Actual Forest Cover

According to the 1993 estimates, the total recorded forest area in Tamil Nadu was about 22,525 square kilometres, of which 17,726 square kilometres or 78.69% was classified as under actual forest cover.[17] In 1997, when the recorded forest area had marginally increased to 22,628 square kilometres, only 17,064 square kilometres or about 75.41% was actually covered by forest.[18] Unmistakably, the actual area under forest cover shrunk over this period. The Report on Forest Sector of Tamil Nadu by Tata Consultancy Services (1995) indicates that about 7000 square kilometres of forest area out of the total 22,699 square kilometres is in a precariously degraded condition.[19]

Table 4.3 Trends in livestock population in Tamil Nadu, 1951–1998

Years	*Cattle*	*Buffaloes*	*Sheep*	*Goats*	*Others*	*Total*
1951	10,161	2285	7926	4013	539	24,924
1956	9698	2041	7042	3757	626	23,764
1961	10,826	2594	7160	3429	629	24,638
1966	10,939	2724	6621	3771	594	24,569
1974	10,572	2853	5393	3954	661	23,433
1977	10,801	3078	5289	4202	776	24,146
1982	10,366	3212	5537	5246	1826	26,187
1989	9353	3128	5880	5910	2085	26,366
1994	9096	2931	5612	5865	2175	25,679
1998	9603	2899	5539	5392	4982	28,415

Source: Tamil Nadu economic appraisal (various years)

Livestock Population

Sustaining livestock largely depends on the accessibility of forest and grazing lands. The post-colonial government took away barren and uncultivable lands along with cultivable waste and pastoral lands for afforestation measures, which led to a scarce resource base for livestock feed. In addition, cattle herds were restricted from entering the reserve forests, with grazing fees and other curbs being introduced, and sheep were banned. These restrictive measures, though intended to preserve forest cover, have resulted in the depletion of cattle wealth (Table 4.3).

It becomes clear that land use patterns in Tamil Nadu have changed, with the changes favourably disposed towards non-forest and other non-agricultural purposes. This poses a serious problem. There has also been a sharp fall in the area classified as common property lands other than forest, a prime reason for the depletion of livestock resources in the same period. Interestingly, though the extent of the net irrigated land has gone up during the Green Revolution period, it has declined in proportion to fallow land, which has increased remarkably after the 1960s.

State Policy on Afforestation

The afforestation programmes carried out so far have not really been intended to increase the area under forest cover but to meet the requirements of the forest-based industries and the fuel-wood demands of local people.[20] In the First Five-Year Plan, the focus of the forest policy was on the resumption and rehabilitation of Ex-zamin as well as the *panchayat* forests and the formation of fuel and matchwood plantations.[21] Under this, plantations were raised on 0.15 lakh hectares of land. The thrust during the Second Five-Year Plan was shifted to the denuded and degraded forests. Further, the priorities included raising the number of industrial and commercial species and conserving the wildlife. Under this plan, about 0.78 lakh hectares of cashew, casuarinas and teak plantations were raised.[22] The Third Five-Year Plan also emphasised the long-term requirements of the state for timber, fuel and raw materials. Under this plan, farm forestry projects with quick growing species such as bamboo and eucalyptus were launched to meet the requirements of the pulp and paper industries. In addition to this, economic plantations such as blue gum, rubber, wattle and bamboo were also raised, which accounted for 0.76 lakh hectares during this period.[23] There was no major change in the Fourth Five-Year Plan, which underlined the need to have more economic plantations and quick growing species for the uninterrupted supply of raw materials to forest-based industries.[24] Such plantations were raised over about 0.68 lakh hectares during this period. The Fifth Five-Year Plan envisaged an increase in production of raw materials for forest-based industries, and fuel plantations to meet the firewood demand from urban areas (Table 4.4).[25]

The Seventh Five-Year Plan gave a little more importance to conservation while emphasising commercialisation. The plan document says that the thrust was 'to increase the green cover in the state through protection, conservation, development, production and watershed management in order to increase the resources of fuel, fodder, small timbers and to raise the productivity of industrial wood plantations and minor forest produce'.[26] With this objective, 1.80 lakh hectares of land was brought under forest cover. The Eighth Five-Year Plan exhibited a genuine concern for conservation, laying emphasis on afforestation, regeneration of degraded forests and the effective prevention of deforestation.[27] In addition to these, the unauthorised and illegal felling of trees for industrial use and fuel as well as other domestic consumption was restricted.[28] Until the early 1980s, the forest policies initiated by the state have given greater importance to the supply of raw materials to industry and fuel-wood for domestic consumption. Though conservation has gained due attention in later plans, the thrust towards commercialisation was never completely withdrawn. In other words, the conservation of forest resources has never been a prime agenda of the state.

Table 4.4 Plan outlay and expenditure and the area covered under the afforestation measures

Plan period	*Approved plan outlay (Hectare in lakhs)*	*Actual plan expenditure (Hectare in lakhs)*	*Forest expenditure to total outlay (Hectare in lakhs)*	*% of forest expenditure to total plan expenditure*	*Area covered (Hectare in lakhs)*
First plan (1951–1956)	8600	8039	30.43	0.38	0.15
Second plan (1956–1961)	18,619	18,776	147.00	0.78	0.78
Third plan (1961–1966)	34,233	34,715	372.14	1.07	0.76
Annual plans (1966–1969)	26,599	26,618	279.99	1.05	0.39
Fourth plan (1969–1974)	55,169	55,896	574.71	1.03	0.68
Fifth plan (1974–1979)	112,232	116,511	1198.33	1.03	1.09
Annual plans (1979–1980)	71,200	71,113	516.13	0.73	0.41
Sixth plan (1980–1985)	315,000	364,462	5700.00	1.56	2.13
Seventh plan (1985–1990)	575,000	631,744	11,189.00	1.77	1.8
Annual plans (1990–1992)	305,000	331,706	8504.00	2.56	0.85
Eighth plan (1992–1997)	1,020,000	1,401,680	23,128.57	1.65	2.77

Sources: Plan documents (various five year plans) and Statistical Hand Book 2002

During the first decade after independence the government has encouraged afforestation through programmes such as Vana Mahotsav (Forest Fest). This continued for a decade but has failed to meet its purpose.[29] From the 1960s onwards the Forest Department shifted its focus to farm forestry and continued with this up to the early 1980s. Afforestation measures were designed until the 1970s so as to increase revenue from forests and infrastructure facilities were initiated to extract more and more forest produce.[30] During the 1960s and 1970s, farm forestry was taken up on wastelands outside the departmental forests, but this turned out to be mono-cultivation, especially of the eucalyptus. Consequently, the scheme failed because of eucalyptus's ecological unviability; moreover, it was not profitable.[31] During the 1970s the extension forestry project was implemented. The Social Forestry Project (SFP), introduced in 1982 with the support of central government and international agencies such as the Swedish International Development Authority (SIDA), primarily focused on the plains and as such the forest-dwellers were not included.[32] Implemented with the objective of involving villagers in afforestation activities, the SFP continued until 1995/1996. Under this programme, the afforestation of community lands in rural areas was taken up on a massive scale with the specific goal of meeting local needs for bamboo, timber, fodder, grass and so on. Between 1981/1982 and 1995/1996, Rs.142.40 crore was spent on this project and about 12.65 lakh hectares were covered.[33] In 1991/1992, the Tamil Nadu Agricultural Development Project (TNADP) was initiated to cover additional areas under forestry programmes with assistance from the World Bank. Under this scheme, Rs.57.68 crore was spent to cover about 28,798 hectares of land, consisting of eco-restoration zones, production zones and interface zones, between 1991/1992 and 1997/1998.[34]

This was followed by the Tamil Nadu Afforestation Project, in 1997/1998 which had financial support from the Japanese Overseas Economic Corporation Fund (OECF). This continued until the close of the twentieth century. The scheme's major objective was to ensure the participation of local people in planning, execution, benefit sharing, monitoring and evaluation under the Joint Forest Management (JFM). Through this scheme, 999 Village Forest Councils (VFC) with a strong representation of 3.59 lakh people were formed to restore about 3.75 lakh hectares of degraded forests.[35] The afforestation initiatives carried out under various nomenclatures all focused on the plain areas, mainly to sustain the supply of raw materials to industries and fuel-wood to the rural as well as the urban populations of Tamil Nadu.

The state government has established three corporations with the explicit purpose of commercial extraction of forest resources: Tamil Nadu Forest Plantation Corporation Ltd, Tiruchirappalli, established in 1974, Tamil Nadu Tea Plantation Corporation Ltd, Nilgiris, started in 1976, and Arasu Rubber Corporation, Ltd, Nagercoil, set up in 1984. The Forest Plantation Corporation Ltd, besides supplying material for wood-based industries, is also engaged in extracting forest resources for domestic and external markets. The Tea Plantation Corporation Ltd was intended to rehabilitate the Sri Lankan repatriates by raising tea plantations in suitable lands in the Nilgiris. Therefore, it is clear that commercial considerations were given more weight in the afforestation programmes that were implemented in the post-independence period in Tamil Nadu.

Trends of Administrative Strength

To protect forest resources from illicit felling and unauthorised occupation, the number of staff in the Forest Department has increased gradually during the post-independence period. The total number of personnel has increased from 5176 in 1967/1968 to 9089 in 1997/1998. The officers alone witnessed a sharp increase from 72 in 1967/1968 to 2165 in 1997/1998. The number of special category staff, working at field level, increased from 4165 to 5575 during the same period (Table 4.5).

With the objective of effectively protecting forest wealth, particularly sandalwood and timber, seven forest protection squads and a mobile patrol team were in operation until 1977. Later on, the number of squads was increased to twelve and three detective flying squads were formed.[36] In order to check the smuggling of forest wealth and illicit activities, a

Table 4.5 Trends of Forest Department administrative staffs strength, 1967/1968–1997/1998

Categories		*1967/1968*	*1977/1978*	*1997/1998*
I	*Superior officers*			
	Principal Chief Conservator of Forests	–	–	3
	Additional Chief Conservator of Forests	–	–	1
	Chief Conservator of Forests	1	2	7
	Conservator of Forests	5	8	20
	Deputy Conservator of Forests	20	35	63
	Assistant Conservator of Forests	37	46	122[a]
	Other Technical Officers	8	16	–
	Others	1	6	–
II	*Special categories*			
	Rangers	200	288	593
	Foresters	821	1047	1323
	Forest Guards	1579	1264	2154
	Forest Watchers	1161	1185	1505
	Others	404	679	
III	*Common categories*	939	1143	3298[b]
Total		5176	5719	9089

Sources: Government of Tamil Nadu, *Synoptic Statistics on Forestry in Tamil Nadu,* 1978, p. 29 and Policy note on Forest Department 2003/2004

[a]Deputation, Training, temporary additional, Training Assistant Conservator of Forests, etc. are included
[b]Others also included

separate Forest Cell Criminal Investigation Department (CID) was established in 1983, together with sixty-nine checkpoints. The number of checkpoints increased to eighty-nine in 1986/1987.

Trends of Forest-Related Offences

Forest-related offences have reportedly decreased in Tamil Nadu in the 1990s when compared with the 1970s. On average, 30,000 cases per annum were recorded in the early 1970s, and this declined to about 20,000 by the late 1990s (Table 4.6). Available data shows that the enhanced administrative strength of the Forest Department has helped to drastically reduce forest-related offences, especially in the last decade. Though the rate of offences has come down, the growth of forest resources has not registered any corresponding increase over the period.

The protection of forest wealth and resources through effective control by increased personnel signified a welcome shift in the state's thinking. Illegal entries into the forests were curtailed to a great extent. However, two things stand out clearly. First, the thrust towards afforestation, with its stress on commercialisation, continued to be a means of providing uninterrupted supply of raw materials to the industries. Hence conservation as a programme of environmental protection received little encouragement. Second, the decline in actual forest cover despite a decrease in forest-related offences thanks to increased administrative strength indirectly indicates that the process has facilitated commercial exploitation.

Diversion of Forestland for Other Purposes

Since independence, both individuals and various government departments have encroached upon forest lands to a large extent. Interestingly, more than half of this land has been used to favour either individuals or private bodies and the rest was meant for governmental development projects. For example, 74,893 hectares of forest land were diverted to other purposes between 1947 and 1977, of which 40,612.50 hectares (54.27%) were for individuals, private bodies and the rehabilitation of non-tribals in the forest areas. About 34,280.71 hectares (45.73%) of forest land were diverted for development programmes such as irrigation projects, hydroelectric projects and railway lines during the same period (Table 4.7). In the early 1970s, a huge number of repatriates from Sri Lanka were rehabilitated in the hill areas of different parts of Tamil Nadu.[37] For instance,

Table 4.6 Trends in the number of offences, 1969/1970 and 1972/1973 to 1996/1997 and 1999/2000

Year	*No. of cases at the beginning of the year*	*New cases during the year*					*No. of cases disposed during the year*			
		Unauthorised felling	*Grazing without permit*	*Other offences*	*Total*	*Total cases of the year*	*Convicted*	*Acquitted*	*Department disposal*	*Total*
1969–1970	75,403	21,321	4771	6174	32,266	107,669	22,670	18,298	40,968	81,936
1970–1971	66,503	24,093	3868	3586	31,547	98,050	20,319	16,326	36,645	73,290
1971–1972	61,003	20,889	4261	4603	29,753	90,756	21,508	13,095	34,603	69,206
1972–1973	56,144	22,036	4507	4766	31,309	87,453	22,110	14,248	36,358	72,716
1996–1997	21,347	9386	1437	7986	18,809	40,156	3684	2553	11,454	17,691
1997–1998	22,465	10,825	1140	8878	20,843	43,308	4838	376	15,328	20,542
1998–1999	22,766	13,292	846	5174	19,312	42,078	2854	248	14,840	17,942
1999–1900	24,136	13,611	3797	4378	21,786	45,922	11,281	3328	11,671	26,280

Source: India Forest Statistics 1969/1970 to 1972/1973 and Administration Report of the Forest Department (various years)

Table 4.7 Forest land diverted for various purposes, 1947–1977

Purposes	*Area (in ha)*	%
Lease to individuals	809.13	1.08
Lease to government department	7002.35	9.35
Lease to private bodies	6286.61	8.40
Area assigned to other government departments	8884.61	11.87
Area assigned to private bodies	24,325.14	32.50
Area released to hydroelectric projects	4696.60	6.28
Area released to special minor irrigation project	1299.19	1.74
Area released to major and medium Irrigation projects	12,286.49	16.35
Area released to railways	102.47	0.14
Rehabilitation purpose	9200.62	12.29
Total	74,893.21	100

Source: Government of Tamil Nadu, *Synoptic Statistics on Forestry in Tamil Nadu*, 1978, p. 20

in 1972/1973, 35.241 square kilometres (3500 hectares) of forest land were demarcated for landless poor people and other development purposes.[38] In 1973/1974, about 70 square kilometres (7000 hectares) were allocated to the same purpose.[39] However, in 1973, the state high power committee under the Chief Minister decided not to divert forest land for non-forest use, particularly to the landless poor.[40]

To prevent this kind of diversion, the Government of India enacted the Forest Conservation Act in 1980. Under this Act, 127 cases involving 3839.33 hectares of forest land were considered until 1991 by central government, in order to decide whether to assign forest lands to various development projects such as the construction of dams. Of these, ninety-six cases concerning 1846.72 hectares of forest land were disposed of favourably. While seventeen of the remaining thirty-one were rejected, fourteen were still under consideration.[41] In 1991/1992, 43 hectares of forest land were diverted to development activities, whereas only 8 hectares were diverted in 1992–1993.[42] Between 1989/1990 and 1993/1994, 108.1 hectares of forest land were assigned for non-forest uses.[43] In 1994/1995, about 9.55 hectares of forest land were diverted for irrigation projects, water supply schemes, electricity transmission lines, approach roads, roads over bridges and repeater stations.[44] In 1995/1996, 1.86 hectares of forest land were taken up for construction of a wireless station, the widening of a river course, water supply, irrigation, construction of primary health centres and a checkpoint.[45] By 31 May 1998, 235 proposals had been received for consideration under the Conservation Act, of which 155 had been approved, nine approved in principle, eleven rejected, one lying with central government, five withdrawn and twenty-four pending with the state government.[46] In addition to this, a sizeable area demarcated for wildlife parks and sanctuaries during the post-independence period. About 134.70 square kilometres, 13.7% of the total forest area, was set apart for wildlife conservation. This included five national parks, eight wildlife sanctuaries and twelve bird sanctuaries, all established during the last three decades by the end of 2005. Tourism promotion and attendant programmes with state patronage have grown by leaps and bounds in the post-independence era, creating pressures on sustenance and preservation of forest cover.

Encroachment is another major problem that continues to elude solution. As at 31 March 1994, 15,423.75 hectares of forest area remained encroached upon.[47] This has increased over the period since, and according to the official figures it was 18,283 hectares by 1998 in Tamil Nadu.[48] Most of these lands were not encroached upon in a strict sense since the government had defined them as forest areas either without consulting the occupants or denying them an opportunity to state their case.

Commercialisation of the Forests

From the late eighteenth century, the colonial government actively pursued a forest policy that facilitated rapid commercialisation. As such, conservation was not accorded due primacy until the late nineteenth century, when the concept of reserve forests was introduced.[49] The idea of having reserve forest was a well-designed strategy to protect and preserve the fauna and flora from being plundered and to conserve the forest wealth. At the same time, the British did not completely give up their commercial motives until the first half of the twentieth century.[50] Here, it is interesting to note that the post-independent policy initiatives were not completely divorced of commercialisation either. In fact, commercial exploitation has continued to guide the policy framework, being actively encouraged in various forms. Invariably, all afforestation projects were aimed at increasing revenue through the supply of raw materials to industries. In the afforestation programmes, the tribals were further marginalised. Bandopadhyay has rightly pointed out that 'these discourses on scientific forestry could neither stop the progressive commercialisation of forestry of various forms and the marginalisation of its resident communities'.[51]

Demand for Forest Products

Growth in industrial as well as non-industrial sectors, which witnessed a substantial leap after independence, spurred a great demand for forest products. Even quantitatively, the demand has been increasing. Some of the industries, which required the continuous supply of forest products, included paper and rayon, railway sleepers, auto and boat building, plywood, sandalwood and other precious oils besides matchwood and tanning. The non-industrial uses include the construction and manufacture of agricultural implements besides fuel-wood, pasture and so on. Industrial development, population growth and urbanisation have created more demand, leading to the depletion of forest wealth. According to official estimates, the demand for forest products, which was at 108 lakh tons in 1970, increased to 113 lakh tons by 1980 (Table 4.8).

Forest-Based Industries

Tamil Nadu can boast prominent forest-based industries such as the Seshasayee Paper Mill in Pallipalayam, the Sun Paper Mill in Tirunelveli, South India Viscose Industries in Coimbatore (rayon manufacturing) and a large cluster of tanning and *beedi* units, as well as other tiny subsidiaries dotting the state. Since independence, the number of forest-based industries has been increasing, and they are dependent upon various kinds of forest produce such as timber, poles, fuel-wood, pulp and matchwood, sal trees and *tendu* leaves, gums and resin. However, rayon and paper are the two major industries that consume the lion's share of forest resources.

The quantity of forest produce supplied to these manufacturers and industries throws some light on the realities on the ground. Several industries in the state are dependent upon forests for the supply of raw materials. The paper industry is the most prominent among them. In 1982/1983 it used a total of 2.33 lakh tons, while a decade later, in 1991/1992, it witnessed a decline, the total supply standing at 2.05 lakh tons. Owing to the stringent norms and restrictions imposed by the government on the extraction of forest produce, this has decreased further. Between 1982/1983 and 1991/1992, about 28.75 lakh tons of forest produce were supplied to different industries (Table 4.9). However, it has been estimated that the required quantity of forest produce was 5.20 lakh tons.

Table 4.8 Estimated of demand of forest products for industrial and non-industrial use in 1970s and 1980s (in tons)

Product		*1970*	*1980*
A	*Industrial use*		
I	Pulpwood bamboo	71,000	80,000
	Paper and boards-wood	23,000	68,000
	Rayon and staple fibre	10,000	18,000
	Wood	100,000	200,000
II	Railway coach building	60,000	5000
	Auto bodies	600	650
	Boat building	3000	3200
III	Railway sleepers	44,000	10,500
IV	Plywood		
	Veneers furniture	5000	6000
	Chip boards	200	1000
V	Match wood	112,000	117,000
VI	Tanning		
	Wattle bark	17,500	17,600
	Wattle extract	1700	1700
	Marobalm	1800	1800
	Konnai bark	100	100
	Avaram	300	300
VII	Cashew shelling (raw nuts)	5000	5000
VIII	Sandalwood	1500	500
IX	Essential oils		
	Lemongian oil	250	250
	Eucalyptus oil	5000	5000
X	Lac	4	20
Total		249,954	224,620
B	*Non-Industrial use*		
I	Constructional Use		
	Timber	107,377	118,114
	Bamboo	16,000	24,000
II	Wood in agriculture		
	Bullock carts	35,886	35,886
	Ploughs	45,640	4540
III	Fuel wood	8,898,500	9,364,500
IV	Pasturage animals	1,530,000	1,530,000
Total		10,633,403	11,077,040
Grand total		10,883,357	11,301,660

Source: Government of Tamil Nadu, *Synoptic Statistics on Forestry in Tamil Nadu*, 1978, pp. 31–32

Table 4.9 Supply of forest produce to forest-based industry in Tamil Nadu: 1982/1983–1996/1997

Year	*Tons*
1982–1983	232,956
1983–1984	222,259
1984–1985	254,131
1985–1986	196,039
1986–1987	217,200
1987–1988	192,923
1988–1989	209,950
1989–1990	194,000
1990–1991	196,563
1991–1992	204,982
1992–1993	185,867
1993–1994	187,084
1994–1995	169,047
1995–1996	119,825
1996–1997	92,003

Source: Tamil Nadu: an economic appraisal (various years)

The number of paper industries has increased over the years since independence, and at the beginning of the 1990s there were twenty-one paper mills with an installed capacity of 2.64 lakh tons. In 2000, there were thirty-one major paper mills with a capacity of 6.39 lakh tons per annum.[52] Of these, nine units had a capacity of 5000 tons per annum, six plants with 5001–10,000 tons, ten with 10,001–20,000 tons, two with 20,001–33,000 tons and four with 33,000 tons and above.[53]

The raw materials to these industries were supplied at a very low rate with a view to sustain industrial growth. In other words, this subsidised supply did nothing but encourage the depletion of forest wealth. Until 1990/1991, the price of raw materials supplied to the industries was very low. For instance, the price per ton was Rs.330 for bamboo, Rs.450 for Eucalyptus Hybrid wood, Rs.925 for bluegum wood, Rs.800 for Eucalyptus Grandis wood, Rs.600 for wattle wood and Rs.2200 for wattle bark. However, the price structure was rationalised in 1995/1996 and in the subsequent years. Between 1990/1991 and 1996/1997, there was a two- to threefold increase in the prices of raw materials (Table 4.10). Until the last decade of the twentieth century, the state continued to supply raw materials to industries at subsidised rates, with a view to boosting industrial development.

Table 4.10 Trend of raw material price for industries (Rs./tons)

Species	*1990/1991*	*1995/1996*	*1996/1997*
Bamboo	330	570	1140
Bluegum	925	1570	2748
Eucalyptus Grandis	800	1360	2380
Debarked wattle wood	600	1020	1785
Eucalyptus Hybrid	450	770	1348
Wattle bark	2200	3375	4725
Pinas patula	660	–	–

Source: Policy note on Forest Department (various years)

Sandalwood

Sandalwood trees are largely found in the hilly tracts of North Arcot, Salem, Dharmapuri and Coimbatore districts of Tamil Nadu. The Nilgiris, South Arcot, Tiruchirapally, Ramanathapuram and Tirunelveli districts also account for this precious tree species. Since the late eighteenth century, sandalwood has been cleared on a massive scale from the forests for commercial purposes, even being exported from different parts of the Madras Presidency.[54] Indeed, it was one of the major sources of revenue for the Forest Department, accounting for about 40% of revenue on its own.

Interestingly, the rate of sandalwood felling has increased over the period, particularly since independence, and it was very high during the second half of the twentieth century. Between 1956/1957 and 1999/2000, 63,297.5 tons of sandalwood worth Rs.38,846.28 lakh were cleared from the forest (Table 4.11). The state has taken up the planting of sandalwood saplings as part of it afforestation measures. Until the 1980s, sandalwood was sold only by the state. In 1983, the government established a sandalwood oil factory at Thekkupattu village of Vaniambadi taluk in North Arcot district. The installed capacity of the plant is 400–500 tons of sandalwood per annum with an output of 22 tons of oil extraction.[55]

While the state was engaged in the sale of sandalwood, illicit felling by smugglers and the forest mafia, who had understandings with the law enforcement agencies, was rampant.[56] The number of offences related to the illicit felling of sandalwood saw a phenomenal rise during the last quarter of the twentieth century. Between 1982/1983 and 1997/1998, about 47,959 instances of illegal felling were detected, 11,144.3 tons of

Table 4.11 Quantity and value of sandalwood extracted from the forests in Tamil Nadu, 1956/1957–1999/2000

Years	*Sandalwood (quantity in tons)*	*Sandalwood (value in lakhs)*	*Years*	*Sandalwood (quantity in tons)*	*Sandalwood (value in lakhs)*
1956–1957	967	68.60	1978–1979	4693	384.10
1957–1958	1162	62.51	1979–1980	4353	364.20
1958–1959	1183	51.40	1980–1981	2411	445.64
1959–1960	1187	58.45	1981–1982	1755	461.13
1960–1961	1882	76.73	1982–1983	2123	658.20
1961–1962	1761	112.96	1983–1984	2660	691.55
1962–1963	1927	81.92	1984–1985	1854	811.44
1963–1964	1385	80.07	1985–1986	1170	1331.81
1964–1965	1400	92.09	1986–1987	1992	1501.00
1965–1966	1496	72.59	1987–1988	1691	1901.90
1966–1967	1221	117.90	1988–1989	1032	1196.50
1967–1968	1082	117.66	1989–1990	968	1273.40
1968–1969	1165	120.23	1990–1991	643	1283.80
1969–1970	1302	115.31	1991–1992	1012	1259.79
1970–1971	1415	122.79	1992–1993	N.A	848.34
1971–1972	1465	122.06	1993–1994	N.A	1315.69
1972–1973	1390	133.71	1994–1995	N.A	2824.00
1973–1974	1370	282.19	1995–1996	N.A	1200.00
1974–1975	1445	372.81	1996–1997	N.A	2636.00
1975–1976	1473	241.91	1997–1998	N.A	1092.88
1976–1977	2110	573.24	1998–1999	N.A	2718.92
1977–1978	2822	404.76	1999–2000	2330.5	9164.00

Sources: Tamil Nadu Economic Appraisal and Policy Note on Forest Department (various years)

N.A not available

sandalwood worth of Rs.23,124.81 lakh were seized and 12,080 vehicles used for smuggling were confiscated in Tamil Nadu (Table 4.12). Illicit felling gained momentum during the early 1990s, with the main reason attributed to a hike in the price of sandalwood from Rs.0.40 lakh per ton in 1960 to Rs.2 lakh in the 1980s, which further went up to Rs.9 lakh in the 1990s. In 1989–1990, 60 tons of sandalwood were seized from the notorious sandalwood smuggler Veerappan and his associates, following the direction of the Chief Minister.[57] Such huge quantities could not be cleared from the forests without the collusion with Forest Department staff. Disciplinary proceedings were initiated against those personnel involved. From 1980 to 1982, 3024 cases were registered against Forest

Table 4.12 Details of offences concerning the illicit felling of sandalwood and other scheduled timbers, 1982–2000

Year	*No. of offences*	*Quantity (in tons)*	*Value (rupees in lakhs)*	*Vehicles confiscated*
1982/1983	384	46.53	19.48	42
1983/1984	417	61.53	16.43	21
1984/1985	349	60.60	21.12	28
1985/1986	670	75.60	40.99	32
1986/1987	457	10.04	40.60	40
1987/1988	N.A	N.A	N.A	N.A
1988/1989	541	113.00	111.33	106
1989/1990	649	184.00	192.83	168
1990/1991	2385	734.00	1147	N.A
1991/1992	2292	768.00	1330.00	712
1992/1993	1799	549.00	958.00	497
1993/1994	3544	843.00	1475.00	661
1994/1995	2817	533.00	933.00	560
1995/1996	2532	523.00	1083.00	531
1996/1997	2760	777.00	1894.34	693
1997/1998	2622	603.00	1962.61	709
1998/1999	2101	N.A	816.00	462
1999/1900	1135	N.A	513.00	164
1990–1991	2385	734.00	1147.08	2385
1991–1992	2046	681.00	1177.00	618
1992–1993	1799	549.00	958.00	497
1993–1994	3544	843.00	1475.00	661
1994–1995	2817	553.00	933.00	560
1995–1996	2532	523.00	1083.00	531
1996–1997	2760	777.00	1894.00	693
1997–1998	2622	603.00	1903.00	709

Sources: Forest Department Administration Report (various years)

N.A not available

Department personnel on disciplinary grounds for negligence in protecting forest wealth, and a total of 1149 persons were punished.[58] During 1985/1986, 106 forest officials and staffs were placed under suspension for various lapses, while charge-sheets have been issued against 1531 officials.[59] This makes it clear that there exist understandings between the forest, revenue and police departments and politicians/mafia, facilitating the illicit smuggling of sandalwood. In fact, in 1995 the Madras High Court suggested to the state government that an Inquiry Commission should be commissioned to find out the details of the links between politicians/mafia and government institutions.[60] A major network that smuggled sandalwood from Tamil Nadu to sandalwood oil factories in Kerala was in operation from the early 1980s.[61] This was exposed by investigations into the procurement processes of some sandalwood oil factories, and criminal cases have been filed in the appropriate courts.[62]

The price of sandalwood shot up remarkably in the last quarter of the twentieth century. In Tamil Nadu, the rate, which was Rs.4005 per ton in 1957, had increased to Rs.10,874 in 1967 and 47,830 in 1977.[63] Since the 1980s, it has witnessed another sharp increase: it was Rs.31,000 per ton in 1980, Rs.78,000 in 1987 and Rs.160,000 in 1990 (see Table 4.13). In 2002, the best quality sandalwood fetched a price not less than Rs.9 lakh per ton.[64] Hence, smuggling continued to thrive on a large scale, with prices skyrocketing in the market. Here, mention should be made again of the notorious poacher and sandalwood smuggler Veerappan, who killed a number of elephants and cut down a large number of sandalwood trees in the north-western parts of Tamil Nadu in the last two decades of the twentieth century. According to media reports he killed over 2000 elephants for ivory and felled sandalwood trees worth Rs.100 crore.[65] In 1989/1990, about 60 tons of sandalwood were seized from the Veerappan gang and eighty-three persons were arrested.[66]

Table 4.13 Trends of Sandalwood prices, 1900–1990

Year	*Rs./Ton*
1900	365
1933	1000
1965	6000
1970	10,000
1980	31,000
1990	200,000
2000	900,000

Source: Rai, Shobha (1990) 'Status and cultivation of Sandalwood in India', Symposium of Sandalwood in the Pacific held at Honolulu, Hawaii, 9–11 April. Actually taken from Ramanathan, C., 1997. Indian Sandalwood Trade. In TED Case Studies: Sandalwood Case. www.american.edu/projects/mandala/TED/sandalwood.htm and Policy note on Forest Department (various years)

Export of Sandalwood Oil

Sandalwood has not only provided revenue for the state but has also brought in precious foreign exchange. About 618.8 tons of sandalwood oil worth Rs.1354 lakh were exported from Tamil Nadu between 1961/1962 and 1982/1983 (Table 4.14). In view of this great export potential, the state government established sandalwood oil extraction factories in the late 1980s. The neighbouring states of Kerala and Karnataka having similar factories, illicit felling was encouraged in the north-western forests of Tamil Nadu, which are contiguous to these two states: this makes illicit felling and smuggling much easier.

Sandalwood oil extraction industries in the border areas of Kerala are procuring smuggled sandalwood from Tamil Nadu. According to the Divisional Forest Officer (DFO), Palakkad: 'It has been reported that there were 21 factories distilling sandalwood oil on the Karnataka and Tamil Nadu border area. On an average each factory consumes 170 tons of sandalwood per year from which it produces six tons of sandalwood oil. Only a small portion of the wood is being harvested officially while the rest comes unofficially from Tamil Nadu and Karnataka.'[67]

Sandalwood oil, known for its fragrance and multiple uses ranging from cosmetics to healthcare products, has a global market and is being exported

Table 4.14 Sandalwood oil exports from Tamil Nadu, 1961/1962–1982/1983

Years	*Quantity (in tons)*	*Value (Rs. in lakhs)*	*Years*	*Quantity (in tons)*	*Value (Rs. in lakhs)*
1961–1962	47	100	1972–1973	96	227
1962–1963	24	44	1973–1974	122	230
1963–1964	N.A	N.A	1974–1975	46	420
1964–1965	52	79	1975–1976	11	59
1965–1966	39	60	1976–1977	0.1	16
1966–1967	58	130	1977–1978	1.7	20
1967–1968	61	148	1978–1979	N.A	N.A
1968–1969	87	255	1979–1980	73	40
1969–1970	76	185	1980–1981	0.2	2
1970–1971	62	150	1981–1982	6.5	20
1971–1972	83	184	1982–1983	8.3	6

Source: Tamil Nadu: an economic appraisal (various years)

N.A not available

to various countries. The quantity exported has declined since the early 1980s owing to stringent procedures put in place by the government. For example, the total sandalwood oil export from the country was only 243 tons between 1987/1988 and 1992/1993.[68] The full breakdown for the period is as follows: 1987/1988 1939 tons; 1988/1989 1926 tons; 1989/1990 1934 tons; 1990/1991 1937 tons; 1991/1992 1965 tons and 1992/1993 1942 tons.[69] The government of Tamil Nadu continued to ask central government to remove the restrictions, a clear indication of its fixed mindset regarding the commercialisation of forest resources.

Revenue from the Forest

From the late eighteenth century, forest resources have been extracted for commercial purposes in different parts of the Madras Presidency by the colonial government.[70] In fact, the same commercial forest policy has continued to persist with minor cosmetic changes until the end of colonial rule. Although post-colonial governments have launched several afforestation programmes, commercialisation and attendant revenue generation have been given the greater policy thrust. In other words, while bringing additional areas into the ambit of forest cover, the government has never given up on the constant extraction of resources. Instead of conservation, meeting future commercial needs was the real motive behind the government in annexing large tracts of forest.

Major forest products such as timber, sandalwood, firewood and other minor forest produce such as bamboo and cane, fibres and flosses have been extracted on a massive scale. For instance, between 1956/1957 and 1985/1986, 737,762 tons of timber worth Rs.3415.47 lakh, 52,459.96 tons of sandalwood worth Rs.7342.34 lakh and 6,463,208 tons of firewood worth Rs.2606.58 lakh have been extracted. In addition to this, bamboo and cane worth Rs.516.39 lakh and minor forest produce such as fibres and flosses worth Rs.4208.31 lakh were also extracted during the same period (Table 4.15).

Even after the 1980s, revenue from various kinds of forest produce increased progressively in Tamil Nadu. For instance, the revenue from various forest produce was about Rs.2773 lakh in 1985/1986 and this increased to Rs.7852 lakh in 1999/2000. Between 1985/1986 and 1999/2000, about Rs.65,724.84 lakh worth of forest produce was extracted, of which Rs.26,005.62 lakhs worth of sandalwood and

Table 4.15 Revenue from forest produce in Tamil Nadu, 1956/1957–1984/1985

Years	*Major forest produce*						*Minor forest produce*	
	Timber		*Sandalwood*		*Firewood*		*Bamboos and cane (value in Rs. lakh)*	*Fibres and flosses (value in Rs. lakh)*
	Qty (in tons)	*Value (in Rs. lakh)*	*Qty (in tons)*	*Value (in Rs. lakh)*	*Qty (in tons)*	*Value (in Rs. lakh)*		
1956/1957	23,486.00	56.83	967.00	68.60	105,940.00	24.14	N.A	N.A
1957/1958	19,180.00	18.03	1162.00	62.51	202,500.00	31.00	7.00	5.53
1958/1959	14,900.00	19.70	1183.00	51.40	337,800.00	32.41	7.83	7.35
1959/1960	26,017.00	27.11	1187.00	58.45	106,530.00	41.40	8.77	8.37
1960/1961	27,355.00	32.97	1882.00	76.73	142,185.00	51.12	8.83	13.74
1961/1962	37,728.00	36.91	1761.00	112.96	84,329.00	58.11	12.28	16.86
1962/1963	44,872.00	31.81	1927.00	81.92	283,775.00	63.74	11.41	17.23
1963/1964	51,091.00	56.53	1385.00	80.07	278,928.00	59.14	16.42	14.96
1964/1965	38,340.00	74.76	1400.00	92.09	343,092.00	61.02	55.29	19.23
1965/1966	34,563.00	65.60	1496.00	72.59	446,766.00	58.52	13.21	14.63
1966/1967	32,589.00	56.49	1221.00	117.90	552,817.00	67.29	21.01	21.29
1967/1968	34,059.00	62.89	1082.00	117.66	377,417.00	18.51	18.17	27.15
1968/1969	42,283.00	43.73	1165.00	120.23	332,000.00	80.40	17.48	30.76
1969/1970	38,682.00	16.26	1302.00	115.31	332,235.00	67.88	15.96	28.96
1970/1971	41,699.00	44.80	1415.00	122.79	278,320.00	83.65	29.45	33.16
1971/1972	36,477.00	85.45	1465.00	122.06	428,306.00	96.98	28.86	75.57
1972/1973	28,440.00	110.52	1390.00	133.71	311,385.00	101.00	24.52	58.05
1973/1974	22,568.00	108.26	1370.00	282.19	224,232.00	107.91	26.22	207.30
1974/1975	20,355.00	125.22	1445.00	372.81	315,670.00	253.78	34.80	126.08
1975/1976	27,514.00	145.76	1473.00	241.91	271,633.00	289.06	33.24	154.46
1976/1977	20,524.00	200.42	2110.00	573.24	206,184.00	198.89	25.01	259.73
1977/1978	26,366.00	126.66	2822.00	404.76	176,040.00	222.11	22.19	320.53
1978/1979	10,422.00	371.12	4693.00	384.10	98,553.00	118.26	27.71	478.08

1979/1980	14,853.00	234.57	4353.00	364.20	56,686.00	72.30	17.21	697.02
1980/1981	8558.00	116.60	2411.00	445.64	40,208.00	93.62	22.79	601.76
1981/1982	3698.00	133.84	1755.00	505.32	40,632.00	73.06	10.73	970.51
1982/1983	4483.00	133.13	2123.00	658.20	66,104.00	127.49	N.A	N.A
1983/1984	4774.00	534.20	2660.00	691.55	18,665.00	39.99	N.A	N.A
1984/1985	1886.00	345.30	1854.00	811.44	4276.00	13.80	N.A	N.A
Total	737,762.00	3415.47	52,459.00	7342.34	6,463,208.00	2606.58	516.39	4208.31

Source: Tamil Nadu: an economic appraisal (various years)

N.A not available

Rs.15,477.74 lakh worth of raw materials were supplied to industry, and Rs.2150.05 lakh worth of non-wood forest produce and Rs.22,246.43 lakh worth of timber, fuel, bamboo, cashew, casuarinas, softwood and farm forest plantations were extracted (Table 4.16). Admittedly, the central government restrictions on sandalwood export has led to a collapse in prices. Coupled with this, the fall in sandalwood prices on the global market has had a direct impact on the domestic market, as witnessed in the declining trends in sandalwood auction in Tamil Nadu during the 1980s.[71] In April 1980, the Ministry of Commerce, Government of India announced a total ban on the export of sandalwood in log, sawn and billet forms, so as to meet the requirements of domestic consumption.[72] Consequently the price decreased to Rs.16,934 per ton in 1998.

Table 4.16 Revenue from forest produce in Tamil Nadu, 1985/1986–1999/2000 (Rs. in lakhs)

Year	*Sandalwood*	*Supply of raw materials to industries*	*Non-wood forest produce*	*Timber, fuel, bamboo, cashew, casuarina, softwood and farm forest plantations and others*	*Total*
1985/1986	1338.00	873.00	95.00	622.00	2773.00
1986/1987	1400.00	618.00	98.00	733.00	2849.00
1987/1988	1560.00	734.00	77.00	1373.00	3744.00
1988/1989	1352.00	894.00	77.00	1053.00	3376.00
1989/1990	1360.00	1113.00	29.00	1166.00	3668.00
1990/1991	1500.00	1357.00	340.00	1014.00	4211.00
1991/1992	1259.79	2083.32	163.48	1004.86	4511.45
1992/1993	848.34	2041.56	153.81	1034.96	4078.67
1993/1994	1315.69	1985.35	125.63	1288.98	4715.65
1994/1995	2824.00	1986.00	156.00	1281.00	6247.00
1995/1996	1200.00	990.00	139.00	2314.00	4643.00
1996/1997	2636.00	183.00	153.00	1260.96	4232.96
1997/1998	1092.88	223.95	227.57	1252.65	2797.05
1998/1999	2718.92	195.56	162.06	2949.02	6025.56
1999/1900	3600.00	200.00	153.50	3899.00	7852.50
Total	26,005.62	15,477.74	2150.05	22,246.43	65,724.84

Sources: Tamil Nadu: an economic appraisal (various years), Policy note on Forest Department (various years)

Note: Figures for 1989/1990 Budget Revised Estimate and 1990/1991 Proposed Budget Estimate

Trends of Income and Expenditure

A perusal of the income and expenditure of the Forest Department will furnish us with a macro-view of the policy thrusts of the post-colonial governments. It will help to ascertain how far conservation and environmental protection has been accorded priority, corresponding to the emphasis laid on commercialisation. The data clearly proves that the Forest Department received more revenue from the forest than expenditure until the 1980s. In other words, the state government during the first three decades of independence encouraged revenue generation through commercialisation. The trend changed after the 1980s, but this does not necessarily mean that the focus of the Forest Department has taken an U turn (Table 4.17). Expenditure being more than income cannot be relied upon as an indicator of renewed efforts aimed at conservation. Since the 1980s, the Forest Department has pumped huge amounts of money into social forestry programmes, the Tamil Nadu Agricultural Development Project and the Tamil Nadu Afforestation Project, to bring additional areas under green cover. Interestingly, all these programmes were supported by external agencies such as SIDA, the World Bank and the OECF, and mainly focused on areas other than the forest.

Conservation Through Different Forest Acts and Rules

In addition to the Tamil Nadu Forest Act of 1882, which is still in force, the post-colonial governments have enacted several Forest Acts and Rules to preserve and protect forest resources from being plundered. They are the Tamil Nadu Preservation of Private Forest Act 1949; Tamil Nadu Hill Areas (Preservation of Trees) Act 1955; Forest (Conservation) Act 1980 with amendments made in 1988; Forest (Conservation) Rules 1981; National Forest Policy 1988; Indian Wildlife (Protection) Act 1972; Tamil Nadu Timber Transit Rules 1968; Tamil Nadu Sandalwood Transit Rules 1967; and Tamil Nadu Sandalwood Possession Rules 1970. These Acts and Rules have imposed several restrictions on accessing and using forest resources and on the transport of timber in general and sandalwood in particular.

Under the Forest Act of 1882, the tribals could claim only the right of way, watercourse, pasture and collection of minor forest-produce (Section 10). These were not allowed as a matter of course either wholly or partially (Section 11) and the Forest Department granted them intermittently,

Table 4.17 Trends of revenue and expenditure in forest, 1960/1961–1997/1998 (Rs. in lakhs)

Year	*Revenue*	*Expenditure*	*Net income*	*Year*	*Revenue*	*Expenditure*	*Net income*
1960/1961	199.35	106.57	92.78	1980/1981	1458.00	1458.68	−0.68
1961/1962	252.27	114.51	137.76	1981/1982	1477.24	1976.00	−498.76
1962/1963	222.86	139.13	83.73	1982/1983	1524.39	2606.00	−1081.61
1963/1964	259.03	169.44	89.59	1983/1984	2067.27	2909.00	−841.73
1964/1965	307.62	185.50	122.12	1984/1985	2467.80	3019.00	−551.20
1965/1966	280.78	209.05	71.73	1985/1986	2751.87	3027.35	−275.48
1966/1967	318.86	206.76	112.10	1986/1987	3043.82	2961.25	82.57
1967/1968	338.65	226.90	111.75	1987/1988	3484.52	3950.71	−466.19
1968/1969	336.71	262.73	73.98	1988/1989	3872.00	4168.00	−296.00
1969/1970	335.81	270.98	64.83	1989/1990	4925.02	5107.39	−182.37
1970/1971	366.08	310.03	56.05	1990/1991	4435.18	6318.56	−1883.38
1971/1972	456.99	394.47	62.52	1991/1992	4457.62	6481.59	−2023.97
1972/1973	490.59	427.51	63.08	1992/1993	4452.62	7961.37	−3508.75
1973/1974	735.41	526.33	209.08	1993/1994	5527.00	5419.23	107.77
1974/1975	921.80	574.40	347.40	1994/1995	6481.00	8653.37	−2172.37
1975/1976	1845.72	657.70	1188.02	1995/1996	5796.59	9367.05	−3570.46
1976/1977	1051.59	652.17	399.42	1996/1997	5273.32	9932.13	−4658.81
1977/1978	1051.59	1111.47	1051.59	1997/1998	4366.17	12,188.50	−7822.33
1978/1979	1141.00	925.88	215.12	1998/1999	6400.35	19,205.77	−12,805.42
1979/1980	1088.00	1155.50	1088.00	1999/1900	8096.16	18,395.23	−10,299.07

Source: Government of Tamil Nadu, *Synoptic Statistics on Forestry in Tamil Nadu*, 1978, p. 27, Tamil Nadu: an economic appraisal (various years) and Policy note on Forest Department (various years)

with quantitative restrictions (Section 12). Fresh clearings for cultivation or any other purpose were also restricted (Section 7). If the tribals used the prohibited items, they were liable for imprisonment for a period of six months or a fine of about Rs.500, or both (Section 21). Grazing on forest land was limited and become liable for suspension (Section 22). Grazing cattle in the closed land would attract imprisonment for a month and/or penalty of Rs.200 (Section 28). Whoever infringed any rules made under Section 26 was to be punished (Section 28A). According to Section 28A (1), in any case where such an infringement related to any scheduled timber, the punishment would be imprisonment for a term which could extend to five years and a fine which could be up to Rs.20,000. The tribals were also forced to render free services to the Police and the Forest Department officers.

As per Section 36A, no person was allowed to keep more than 5 kilograms of sandalwood without a licence granted by the District Forest Officer, or unless such sandalwood was affixed with a mark and in such a manner prescribed by a Forest Officer. Further, this act provided for the arrest and seizure of property including implements, vehicles and cattle that were used to commit any offence. The Act included a deterrent against unnecessary seizure or vexatious arrest. The use of timber for housing and to make agricultural implements was curbed under the Tamil Nadu Hill Areas (Preservation of Trees) Act 1955. Section 3(2) of this act reads: 'no person shall cut or otherwise damage, or cause to be cut or damaged, the branch of any trees'. These restrictions were applicable even on the steep slopes of the hills (Section 4(1). Any offender was liable for punishment with imprisonment of up to one year or a penalty of up to Rs.5000 or both (Section–7). After independence, the tribals' rights to cut down trees even for their basic necessities were restricted further.

The Madras Forest (Amendment) Act 1979 increased the penalty for offences committed under the Madras Forest Act (1882). For example, the punishment for using prohibited forest items was raised from six months' to three years' imprisonment or from Rs.500 to Rs.10,000 or both. Later, the Tamil Nadu Forest (Amendment) Act 1992 hiked the penalty amount for offences related to scheduled timber from Rs.10,000 to Rs.20,000 and the imprisonment period from three to five years.[73] The rights of tribals in the forest were thus curtailed systematically over this period, leading to the collapse of the forest-oriented tribal economy.

Restriction on Timber Transit

The government has framed several Rules in order to regulate the transport of timber from one region to another. The Tamil Nadu Timber Transit Rules 1968 restricted the transport of timber without a valid permit. Further, the permit for import as well as export of timber has to be produced at any notified checking station.

Sandalwood Protection

In addition to the general acts, the Tamil Nadu Government has brought out specific rules viz., Sandalwood Transit Rules 1967 and Sandalwood Possession Rules 1970 to protect sandalwood resources. These rules were made for the exercise of powers conferred by sections 35, 36, 36A, 36B and 36D of the Tamil Nadu Forest Act 1882 (Tamil Nadu Act V 1882), to prevent the illicit felling and smuggling of sandalwood.

Besides restricting any person from carrying more than 5 kilograms of sandalwood, the Sandalwood Transit Rules 1967 were provided by the District Forest Officer to maintain an approved list of companies or individuals engaged in the distillation of sandalwood and other byproducts. The export/import and transport of sandalwood without a valid permit was forbidden. In the case of sandalwood purchased at the government sales depot in the States of Andhra Pradesh, Karnataka or Kerala and exported to another state, the permit should be produced from the authority concerning the respective states. In case of imports, permit has to be obtained on payment of a price fixed by the Chief Conservator of Forest from time to time (Section 4(3)). The subsequent movement of imported sandalwood as well as the transport of all local sandalwood should be according to permission from the District Forest Officer on payment (Section 4(4)). Sandalwood transport is also restricted at night unless there is clearance at the first checking station in a district. According to the Tamil Nadu Sandalwood Possession Rules 1970 no person should keep a stock of sandalwood in excess of 5 kilograms without a licence.

Restriction to Divert the Forest Land

Concerned about the need to protect the dwindling forest cover in the country, the Government of India came up with several statutes and rules in the last quarter of the twentieth century. The Forest (Conservation) Act

1980 expressively prohibited state governments from transferring the reserve forests for any purpose without the approval of the Government of India. As per the modified Act in 1988, the states should not assign by way of lease or otherwise any forest land to any private person or authority, corporation, agency or any other organisation not owned, managed or controlled by the government.

Tribal Development Policies and Their Impacts

According to the 2001 Census, the population of Scheduled Tribes in Tamil Nadu is 6.51 lakh, constituting 1.04% of the total population (62,405,679). About 84.62% of them are living in the rural areas. Of the total thirty-six tribal communities in the state,[74] six are classified as Primitive Tribal Groups (PTG).[75] Until 1978 when the Tribal Sub-Plan (TSP) was introduced, the government did not have a major strategy or a development programme for the socio-economic development of the tribals. In fact, budgetary allocation also remained very low and no specific schemes were implemented during the first decade. Only in 1961 were two separate Block Development Offices (BDOs) set up in the Shervaroy and Kolli hills, which were converted into Tribal Development Blocks (TDBs) in 1963. During the Fifth Five-Year Plan, the concept of TSP was introduced, under which nine Integrated Tribal Development Projects (ITDPs) were identified in six districts of north and north-western Tamil Nadu.[76]

With the introduction of the TSP, the state increased the budgetary allocation for the development of tribals. From Rs.2167 lakh (0.59%) in the Sixth Five-Year Plan, the outlay progressively increased to Rs.27,432 lakh (1.10%) by the Ninth Five-Year Plan (Table 4.18). Since the sub-plan period (1978), the hill areas started to gain more attention and infrastructure facilities. A separate Directorate for the Welfare of the Tribals was set up on 1 April 2000. However, these development initiatives failed to have a positive impact on the tribal economy, as there was a visible decline in the proportion of the rate of participation in economic activities. No significant structural transformation has occurred even among the low proportion of workers.[77] Interestingly, there was a sharp fall in the proportion of cultivators to the total workforce in the primary sector (from 68% in 1961 to 43% in 1991) while the category of agricultural labourers registered an alarming increase (about 24% in 1961 to 50% in 1991) (Table 4.19). In addition to this, about 10% of the workers were classified

Table 4.18 Outlay for the tribal development in Tamil Nadu, Plans Two to Nine (Rs. in lakhs)

Plans	*Total outlay*	*Tribal development*	*% of col. (3) to (2)*
Second Five-year plan (1956–1961)	18,776	31.68	0.17
Third Five-year plan (1961–1966)	34,715	41.00	0.12
Fourth Five-year plan (1969–1974)	55,896	76.37	0.14
Fifth Five-year plan (1975–1979)	83,352	83.95	0.10
Sixth Five-year plan (1980–1985)	364,461	2167.00	0.59
Seventh Five-year plan (1985–1990)	631,744	6932.00	1.10
Eighth Five-year plan (1992–1997)	1,401,680	10.621.70	0.76
Ninth Five-Year Plan (1997–2002)	2,500,000	27,432.00	1.10

Source: Five-Year Plan Documents (various plans)

Note: Only the state plan outlay was taken into account

in the category of marginal worker. This is a clear indication that the tribal economy was on a downward slide, with a deterioration having set in owing to large-scale encroachment by the Forest Department and the alienation of tribal lands by non-tribals.

Trends of Landholding Pattern

Without any viable economic transformation, the occupational position of the tribals degenerated to agricultural labourers from cultivators. Not only did the proportion of cultivators dwindle during the post-independence period, even the medium and large farmers declined sharply. On the contrary, the ratio of marginal and small farmers increased in these years. During 1979/1980, marginal landholdings constituted 53.06% of the total holdings and 15% of the total area. In 1995/1996, they had increased to 57.40% of the total holdings and 21.23% of the total area. Small landholdings, which constituted 24.02% of the total holdings and 20.6% of the total area, rose to 26.02% of the total holdings and 29.36% of the total area during the same period (see Table 4.20). The sharp fall in the proportion of semi-medium, medium and large holdings and even their area corresponded to the phenomenon of a very large number of cultivators being reduced to agricultural labourers. This illustrates the deterioration and collapse of the tribal economy in the state.

In fact, the numerous restrictions imposed on accessibility to forest resources and the difficulty of expanding land for cultivation owing to the

Table 4.19 Trend of Scheduled Tribes workforce, 1961–2001

Year	*Total population*	*Total workers*	*% of col. 3 to 2*	*Total primary workers*	*% of col. 5 to 3*	*Cultivators*	*% of col. 7 to 5*	*Agri. labourers*	*% of col. 9 to 5*	*Agri. allied workers*	*% of col. 11 to 5*
1	*2*	*3*	*4*	*5*	*6*	*7*	*8*	*9*	*10*	*11*	*12*
Total											
1961	251,991	143,062	56.77	123,346	86.22	83,289	67.72	29,542	23.95	10,515	8.33
1971	311,515	137,295	44.07	127,771	93.06	61,698	48.29	51,204	40.07	14,869	11.64
1981	458,462	224,497	48.97	200,398	89.27	105,537	52.66	77,572	38.71	17,289	8.62
1991	574,194	281,933	32.51	245,552	87.10	105,345	42.90	123,364	50.24	16,843	6.86
Rural											
1961	237,663	136,457	57.42	120,666	88.43	82,947	68.74	28,249	23.41	9470	7.85
1971	294,379	130,280	44.07	123,706	94.95	61,414	49.65	49,259	39.82	12,933	10.55
1981	415,165	208,746	50.28	194,589	93.22	105,026	53.97	74,775	38.43	14,788	7.60
1991	505,208	258,217	33.84	237,904	92.13	104,869	44.08	118,631	49.87	14,404	6.06
Urban											
1961	14,328	6605	46.10	2680	40.58	342	12.76	1293	48.25	1045	38.99
1971	17,136	7015	40.94	4065	57.95	284	6.99	1945	47.85	1836	45.16
1981	43,297	15,751	36.38	5809	36.88	511	8.80	2797	48.15	2501	43.05
1991	68,986	23,716	22.77	7648	32.25	476	6.22	4733	61.89	2437	31.89

Source: Census of India (Various Censuses)

Note: For 1981 and 1991 Census, total workers include Marginal workers

Table 4.20 Land size holdings of tribes in Tamil Nadu, 1979/1980 and 1990/1991

Size	*Number of holdings*	*% to total holding*	*Area*	*% of total to total area*
Marginal				
1979–1980	25,446	53.06	11,675	14.99
1990–1991	36,703	55.05	16,600	19.15
1995–1996	35,658	57.40	16,767	21.23
Small				
1979–1980	11,520	24.02	16,076	20.64
1990–1991	17,370	26.05	24,275	28.00
1995–1996	16,164	26.02	22,655	29.36
Semi-Medium				
1979–1980	7175	14.96	19,658	25.23
1990–1991	9416	14.12	25,657	29.59
1995–1996	7577	12.20	20,474	26.54
Medium				
1979–1980	3253	6.78	18,920	24.29
1990–1991	2948	4.42	16,476	19.00
1995–1996	2476	3.99	13,647	17.69
Large				
1979–1980	561	1.17	11,577	14.86
1990–1991	240	0.36	3686	4.25
1995–1996	247	0.40	3613	4.68
Total				
1979–1980	47,955	100.00	77,906	100.00
1990–1991	66,677	100.00	86,694	100.00
1995–1996	62,122	100.00	77,156	100.00

Source: World Agricultural Census for the respective years

extension of reserve forests and subsequent land alienation have crippled the tribal economy. Of all the factors, land alienation remains the most serious that the tribal communities have confronted in recent years, especially after the sub-plan period.

A survey undertaken among 2631 tribal households in 1986 to ascertain the intensity of land alienation revealed that 37% of the tribal households had lost their lands to outsiders. On an average each household had lost not less than 3.68 acres of land. Another survey held in nine districts during 1998 showed that of the 3341 tribal households enumerated 1535 had lost about 4647.92 acres of land primarily owing to disposal after the land was mortgaged to meet domestic consumption, debt repayment, social and religious ceremonies/functions, medical expenses and to some extent alcoholism.[78] Though a draft bill on land alienation has been prepared, it is yet to see the light till 2000.

Conclusion

The policy framework of the post-colonial governments in regard to forests was flawed and inherently contradictory, as there was a lack of vision to address environmental and tribal concerns. As such the policies initiated and the programmes implemented failed to bring any desirable results. On the one hand, the state continued to encroach upon forest land for various development purposes and on the other, it offered encouragement to denudation and the plunder of forest resources for commercial gain. The shrinking of the green cover and depletion of forest wealth is a disturbing trend, which will pose an environmental threat to the state in the long run. While restricting the use of forest resources by the tribals, the state has either remained a mute spectator to the looting of timber and other resources or stood by apparently ineffective in controlling contractors/politicians/mafia. Deforestation and denudation caused by the tribals was magnified in order to put curbs on them, but illicit felling and the smuggling of forest wealth, despite stringent laws, raise questions about accountability. Another interesting point is that while providing infrastructure facilities to the tribal areas, the state either overlooked the systemic failures of governance or failed to impose restrictions in order to control non-tribal intruders into the tribal areas. In short, in the post-colonial period there was no paradigmatic shift in the perspective of the state in regard to the hill areas and the tribals concerned. Instead, the pace of disintegration of the traditional forest-oriented tribal economy was accentuated. During the second half of the twentieth century in Tamil Nadu, the policies pursued have neither protected the environment nor promoted the tribal inhabitants, but only paved the way for non-tribals to destroy the environment and the tribal economy.

Notes

1. Government of Tamil Nadu, *State Administrative Report 1989–1990*, 1993, p. 145.
2. For the detailed review see Velayutham Saravanan, 'Colonial Commercial Forest Policy and the Tribal Private Forests in Madras Presidency: 1792–

1881'. *Indian Economic and Social History Review*, 40(4), 2003, pp. 403–407; Velayutham Saravanan, 'Colonialism and Coffee Plantations: Decline of Environment and Tribals in Madras Presidency during the 19th Century', *Indian Economic and Social History Review*, 41(4), 2004, pp. 464–488 and Velayutham Saravanan, 'Agrarian policies in the tribal areas of Madras Presidency during the pre–survey and settlement period, 1792–1872, *Indian Journal of Agricultural Economics*, 65(2), 2010, pp. 261–276.

3. Saravanan, 'Agrarian policies in the tribal areas of Madras Presidency during the pre–survey and settlement period, 1792–1872'.
4. Saravanan, 'Colonial Commercial Forest Policy and the Tribal Private Forests'.
5. Velayutham Saravanan, 'Tribal Revolts in India with Reference to Salem and Baramahal Districts of Madras Presidency during the late 18th Century'. *Artha Vijnana*, 41(1), 1999, pp. 67–81.
6. Saravanan, 'Colonial policy and Tribal Welfare Measures in Madras Presidency'.
7. Saravanan, 'Commercialisation of forest, Environmental Negligence and Alienation of Tribal Rights' pp. 125–146; Saravanan, 'Colonial Commercial Forest Policy and the Tribal Private Forests', 'Colonialism and Coffee Plantations: decline of Environment and Tribals' pp. 464–488; Velayutham Saravanan, 'Economic Exploitation of Forest Resources in South India during Pre–Forest Act Colonial Era, 1793–1882,' *International Forestry Review: The International Journal of Forest Science and Policy*, 10(1), 2008, pp. 65–73.
8. Saravanan, 'Commercial Crops, Alienation of Common Property Resources and Change in Tribal', pp. 298–317; Saravanan, 'Colonialism and Coffee Plantations'.
9. Saravanan, 'Colonial Agrarian Policies in the Tribal Areas of Madras Presidency: 1872–1947'.
10. Velayutham Saravanan, 'Subalterns vs State Institutions: Politicians, State, Forest, Law and Atrocities on Tribals in Tamil Nadu, 1990–2000' *The International Journal of Human Rights*, 15(6), 2010, pp. 948–968; Velayutham Saravanan, 'Terrorising Tribals: Nexus between Police and Forest Department'. *Economic and Political Weekly*, 32(29), 1997, pp. 1789–1790.
11. Velayutham Saravanan, 'Tribal Land Alienation in Madras Presidency during the Colonial Period: 1792–1947'. *Review of Development and Change*, 6(1), 2000, pp. 73–104.
12. Velayutham Saravanan, *Economic Transformation of Tribals in Tamil Nadu Since the Colonial Rule 1792–1991*, unpublished PhD dissertation, Hyderabad: University of Hyderabad, 1994.

13. Saravanan, 'Terrorizing Tribals: Nexus between Police and Forest Department,' pp. 1789–1790; Velayutham Saravanan, 'Tamil Nadu: An Ecological Tragedy'. *Frontier*, 33(3), 2000, pp. 4–6; Ajit Menon and Velayutham Saravanan, 'Displacement and Rehabilitation Policies: The case of Kolli Hills Hydro–electric Project,' *Economic and Political Weekly,* 31(43), 1996, pp. 2854–2855.
14. N.S. Jodha, *Life on the Edge: Sustaining Agriculture and Common Property Resources in Fragile Environment*, Delhi: Oxford University Press, 2001, p. 135.
15. Arun Bandopadhyay, '*Three issues from a CPR management: Village forestry in post-colonial South Asia' in* B.B. Chaudhuri and Arun Bandopadhyay (eds), *Tribes, Forest and Social Formation in Indian History*, New Delhi, Manohar, 2004, p. 212.
16. Government of Tamil Nadu, Policy note on Forest Department 1990–1991, p. 3.
17. Indian Council of Forest Research & Education, *Forestry Statistics India,* Dehra Dun: ICFRE, 1995, p. 26.
18. Indian Council of Forestry Research and Education, *Forestry Statistics of India*. Dehra Dun: ICFRE, 2000.
19. Government of Tamil Nadu, *Ninth Five- Year Plan 1997–2002*, Chennai, 1998, p. 796.
20. Policy note on Forest Department 1973–1974.
21. Government of Tamil Nadu, *Tamil Nadu – An Economic Appraisal 1971–1972*, p. 21.
22. Ibid.
23. Ibid.
24. Ibid., p. 21.
25. *Tamil Nadu – An Economic Appraisal 1972–1973*, p. 29.
26. Government of Tamil Nadu, *Eighth Five–Year Plan 1992–1997*, Madras, 1992, p. 133.
27. Ibid., p. 137.
28. Ibid.
29. Bandopadhyay, 'Three issues from a CPR management', p. 210.
30. *Policy note on Forest Department 1972–1973*, pp. 16–17.
31. Bandopadhyay, 'Three issues from a CPR management' p. 211.
32. Ibid.
33. *Tamil Nadu An Economic Appraisal 1997–1998*, p. 47.
34. *Tamil Nadu An Economic Appraisal 1999–2000*, p. 42.
35. *Tenth Five–Year Plan 1997–2002*, p. 792.
36. *Policy note on Forest Department 1986–1987,* p. 3.
37. *Policy note on Forest Department 1972–1973.*
38. Ibid., p. 2.

39. *Policy note on Forest Department 1973–1974*, p. 1.
40. Ibid., p. 2.
41. *Policy note on Forest Department 1990–1991*, p. 3.
42. *Tamil Nadu: An Economic Appraisal, 1993–1994*, p. 103.
43. Forestry Statistics India, 1995, p. 42.
44. India Forestry Statistics, 1996.
45. Ibid.
46. India Forestry Statistics 2000.
47. Forestry Statistics India, 1995, p. 36.
48. India Forestry Statistics 2000.
49. Saravanan, 'Commercialisation of forest, Environmental Negligence'; Saravanan, 'Colonial Commercial Forest Policy and the Tribal Private'; Saravanan, 'Colonialism and Coffee Plantations.
50. Saravanan, 'Colonialism and Environment: Commercialisation of Forest and Decline of Tribals in Madras Presidency, 1882–1947'.
51. Bandopadhyay, 'Three issues from a CPR management' p. 210.
52. India Forestry Statistics 2000.
53. Ibid.
54. Saravanan, 'Commercialisation of forest, Environmental Negligence'.
55. *Policy note on Forest Department 1985–1986*, p. 9.
56. Saravanan, 'Terrorising Tribals: Nexus between Police and Forest.
57. *Policy note on Forest Department 1990–1991*, p. 54.
58. *Policy note on Forest Department 1983–1984*, p. 4.
59. *Policy note on Forest Department 1986–1987*, p. 5.
60. Saravanan, 'Subalterns vs State Institutions: Politicians, State, Forest, Law and Atrocities on Tribals in Tamil Nadu'.
61. *Policy note on Forest Department 1985–1986*, p. 6.
62. Ibid.
63. Government of Tamil Nadu, *Synoptic Statistics on Forestry in Tamil Nadu*, Office of the Chief Conservator of Forests, Madras, 1978, p. 14.
64. The Hindu dated 22 September 2002 in http://www.thehindu.com/2002/09/22/stories/2002092204570600.htm.
65. The Hindu 2 May 2003.
66. *Policy note on Forest Department 1989–1990*, p. 54.
67. Hindu 13 March 2003.
68. Food and Agriculture Organisation (FAO) Flavours and Fragances of Plant Origin, 1995, in http://www.fao.org/docrep/v5350e/V5350e08.htm.
69. C. Ramanathan, Indian Sandalwood Trade. In TED Case Studies: Sandalwood Case, 1997. www.american.edu/projects/mandala/TED/sandalwood.htm.

70. Saravanan, 'Commercialisation of forest, Environmental Negligence' 'Colonialism and Environment: Commercialisation of Forest and Decline of Tribals in Madras Presidency, 1882–1947'.
71. *Policy note on Forest Department 1979–1980*, p. 7.
72. P*olicy note on Forest Department 1981–1982*, p. 10.
73. Tamil Nadu Forest (Amendment) Act 1992.
74. They are 1. Adiyan 2. Aranandam 3. Eravallan 4. Irular 5. Kadar 6. Kammara (excluding Kanyakumari district and Shenkottah taluk of Tirunelveli district) 7. Kanikaran, Kanikkar (excluding Kanyakumari district and Shenkottah taluk of Tirunelveli district) 8. Kaniyan, Kanyan 9. Kattunayakan 10. Kochu Velan 11. Konda kapus 12. Kondareddis 13. Koraga 14. Kota (excluding Kanyakumari district and Shenkottah Taluk of Tirunelveli district) 15. Kudiyal Melakudi 16. Kurichchan 17. Kurumbas (in the Nilgiris district) 18. Kurumans 19. MahaMalasar 20. Malai Arayan 21. Malai pandaram 22. Malai Vedan 23. Malakkuravan 24. Malasar 25. Malayali (in Dharmapuri, North Arcot, Pudukottai, Salem, South Arcotand Tiruchirappalli districts) 26. Malayakandi 27. Mannan 28. Mudugar, Muduvan 29. Muthuvan 30. Pallayan 31. Palliyan 32. Palliyar 33. Paniyan 34. Sholaga 35. Toda (excluding Kanyakumari district and Shenkottah taluk of Tirunelveli district) 36. Uraly.
75. Toda, Kota, Kurumba, Irular, Pania and Kattunaickan.
76. They are: Kolli hills, Shervaroy hills, Kalrayan hills and Pachamalai hills of Salem district; Kalrayan hills of South Arcot district; and Jawadhi–Elagiri hills of North Arcot district; Pachamalai hills of Tiruchirappalli and Sitteri hills of Dharmapuri district.
77. Saravanan, 'Decline of Tribal Economy in Tamil Nadu: 1947–2000'.
78. E. Karuppaiyan, 'Alienation of Tribal Lands in Tamil Nadu: Panel Data Analysis' *Economic and Political Weekly*, 35(37), 2000, pp. 3344–3348; see also Institute of Techno-Economic Studies. *Indebtedness of Scheduled Tribes in Tamil Nadu*, Madras, 1978.

Check for updates

CHAPTER 5

Environment and Subaltern Struggle, 1990–2000

Introduction

Conservation as a state initiative commenced in the Indian subcontinent in the last quarter of the nineteenth century with the enactment of the Indian Forest Act 1878. In addition to the national effort, different state governments have also brought forth similar Forest Acts. Through these, severe restrictions were imposed on the tribals and on other forest-dwellers. Implementing these Acts was vested with the three core departments of the state administration, the forest, police and revenue departments, but they have failed miserably to arrest the rise in illicit smuggling, resulting in the decline of forest resources, especially timber and sandalwood in the hills of Tamil Nadu. While the timber mafia flourished, with the active connivance of these institutions of the state, the tribals faced harassment for the ongoing smuggling and illicit felling. Instead of acting within the bounds of law, the state institutions moved away from their due responsibilities.

To protect the Scheduled Castes and Tribes, the Scheduled Castes and the Scheduled Tribes (Prevention of Atrocities) Act 1989 was enacted. Government servants were brought under the act for offences related to the negligence of their duties. A public servant committing any offence under

An earlier version of the chapter was published in *The International Journal of Human Rights*, Vol. 15, No. 6, 2011.

© The Author(s) 2018

V. Saravanan, *Environmental History and Tribals in Modern India*, https://doi.org/10.1007/978-981-10-8052-4_5

this section 'shall be punishable with imprisonment for a term which shall not be less than one year' (Sections 3–4). Further, Section 4 says: 'Whoever, being a public servant but not being a member of a Scheduled Caste or a Scheduled Tribe, wilfully neglects his duties required to be performed by him under this Act, shall be punishable with imprisonment for a term which shall not be less than six months but which may extend to one year.' But the law enforcement agencies such as forest, police and revenue department authorities at the district level teamed up with each other and committed atrocities either by fabricating incidents or implicating tribals on false charges. The casualty in all this was the basic human rights of the tribals.

Given the background, this chapter analyses the nexus formed between different departments of the state administration, the forest, police and revenue, in the violating the human rights of the tribal communities in the Dharmapuri district of Tamil Nadu during the last decade of the twentieth century (1990–2000). It also looks into the relationship between the Forest Department and politicians in extracting forest resources. On the one hand, the state has enacted several Forest Acts and rules besides imposing restrictions on the tribals and other forest-dwellers, whose livelihood is dependent upon the jungles; and on the other hand, the functioning of the Forest Department paved the way for powerful politicians/contractors to plunder forest wealth, this despite the emphasis on conservation and protection of the green cover. How the tribals' basic human rights were violated by the different departments of the state administration and how an incident was fabricated together with the judicial process in securing justice is the focus of this chapter. In short, it attempts to analyse how tribal rights were violated by the different departments of the state administration and how the issues raised were addressed by left-oriented parties, a state of affairs that led to judiciary intervention to help the affected tribals in the north-western parts of Tamil Nadu during the last decade of the twentieth century.

Background to the Vachathi Incident

In 1985, the Forest Department filed two cases against two people of Vachathi village, charging them with illicit felling of forest resources under the Scheduled Timber Offences Report (STOR). In 1992 one more case was filed against another tribal. But the suspects were discovered to be absconding. According to the Forest Department, 'as early as on 30–12–91 a raid was organized, on a small scale to detect such sandalwood smuggling activities in and around Vachathi village. There was very stiff resistance from the villagers and the raid party was attacked. In April 1992, the Shervaroys North Range organized a similar raid. Again the Vachathi villagers resorted to attacking the forest staff, preventing the discharge of their legitimate Government duties.'[1] Regarding the absconded persons, the Deputy Conservator of Forests at Harur directed M. Ramamurthy, ranger of Uttangarai, to obtain non-bailable warrants (NBWs) against the accused. In all twenty-one NBWs were issued in the three STOR cases, including fifteen in the 1992 absconding cases.[2] Later, on 20 June 1992, there was a raid on Vachathi village.

Natural Resources and Livelihood

Historically, the tribals in this area as elsewhere led a self-sustaining forest-oriented life, managing their livelihood through the cultivation and collection of forest produce in different parts of Madras Presidency. The Sangam literature, the earliest available literary sources, stands witness to this, providing much information about life in the hills and jungles.[3] Although there is a controversy about the Sangam period, it is widely accepted that it covers a vast time three centuries before and after the beginning of the Christian era. Until colonial intervention, life went on without any major intervention either by the rulers or by non-tribals from the plains.[4] The same trend would have prevailed in Vachathi village, the area focused upon here. Chitteri hill was abundant with sandalwood trees, which increased after the introduction of the reserve forest in the 1880s. There was no threat to this species until the early independence period, but the situation changed in just two decades owing to rampant illicit felling and smuggling not only in this region but in all the hill ranges of Tamil Nadu.

Vachathi village is located about 14 kilometres from the taluk headquarters of Harur in the Dharmapuri district of Tamil Nadu. Surrounded by Chitteri hills on three sides, the village is at an intersection of the Harur and Papireddipatti forest ranges. In the early 1990s, it had a total population of 655 of whom 643 (98.17%) were Malayalis, a major Scheduled Tribe in Tamil Nadu, while the remaining twelve persons (1.83%) belonged to other castes. Before the state stepped in with developmental measures, these people lived in huts that were constructed with locally available materials. The village received attention only after the introduction of the Tribal Sub-Plan (TSP) during the Fifth Five-Year Plan and gained some infrastructure facilities. The total number of households was about 200, with 120 of their houses having been constructed under the Integrated Tribal

Development Project (ITDP). In 1985–1986, about sixty houses were constructed under the Rural Landless Employment Guarantee Programme (RLEGP). In addition to this, three to four *pucca* houses were constructed. According to the 2001 census, the total number of households in Vachathi was 230; the total population was 939 (479 males and 460 females), with 889 tribals, four Scheduled Castes and the remaining forty-six belonging to other communities. Only 371 were literate. Of this population, 484 were classified as workers, of whom 163 were marginal workers. Cultivators numbered 196 and marginal cultivators 157, while 106 were classified as agricultural labourers. This clearly indicates that more than half of the people were non-workers and illiterate, while a very large proportion belonged to the marginal agricultural labourers category.

Agriculture and its allied activities was the mainstay of their livelihood. The total cultivated area of this village was 251.18 hectares and only 21% of the population (190 persons) had landholdings, while the others (79%) were landless labourers. The main crops raised were *ragy*, tapioca, rice and cereals. The collection of minor forest produce and firewood were part of their routine. In addition to this, they engaged in farming as wage labourers.

The two reserve forests, Thombakkal and Pallipatti, that were adjacent to the village grew plenty of sandalwood among other things. An enumeration in the 1830s listed close to 5000 sandalwood trees in the area.[5] During the early nineteenth century, sandalwood was cut down by the colonial government in order to be sold.[6] However, the tribals were not aware of the commercial value of this precious wood. In fact, the market for sandalwood has grown by leaps and bounds only during the last quarter of the twentieth century.[7]

As mentioned earlier, Vachathi received some infrastructural facilities such as roads from the 1980s. Yet the lack of civic amenities was still woeful by the 1990s: there was only one overhead tank with a 33,000 gallon capacity, with eleven tap connections supplemented by water lifted from a well with an electrical pump, which was non-functional for a long time. Nine streetlights were maintained in the village. The tribal residential school could accommodate 100 children, while the Balwadi (day care centre) had a maximum intake of sixty children up to five years of age. The high school was at Pethathampatti 1.5 kilometres away and the higher secondary school at Harur was 14 kilometres from Vachathi village.

Modern health care was out of reach: villagers had to walk up to 8 km to the nearby primary health centre (PHC) at Chinnankuppam. The mobile clinic with a doctor and a nurse was supposed to make regular visits on

Wednesdays, but this seldom occurred.[8] There were 145 ration cardholders who received essential commodities through the fair price shop.[9] In recent years, many Vachathi tribals had started to rear goats, subsidised by government loans, but they had to pay Rs.10 for grazing hybrid goats and Rs.3 for local varieties in the forest. The Forest Department did not record payments with proper receipts and at times seized cattle and poultry.[10] These were the socio-economic characteristics of Vachathi in the early 1990s.

State Institutions and Atrocities Regarding Tribals

Sandalwood was extracted from the forests for export. Even after the introduction of the reserve forest system, particularly after the 1880s, the Forest Department continued with this extraction, but more systematically, until the end of the colonial period.[11] However, illicit felling is not mentioned in the government records. Most probably it would not have attracted the attention of the administration, since prices were very low and this meant the sandalwood processing industries were almost non-existent at this time.[12]

Smugglers and the forest mafia, who had various understandings with the law enforcement machinery, flourished during the last quarter of the twentieth century in different parts of Tamil Nadu.[13] The number of offences related to the illicit felling of sandalwood saw a phenomenal rise. Between 1982–1983 and 1997–1998, about 47,959 instances of illegal felling of sandalwood trees were detected; 11,144.3 tons of sandalwood worth Rs.23,124.81 lakh was seized and 12,080 vehicles used for smuggling were confiscated.[14] Illicit felling gained momentum during the early 1990s, the main reason for this being the hike in the price of sandalwood from Rs.0.40 lakh per ton in 1960 to Rs.2 lakh in the 1980s, this further increasing to Rs.9 lakh in the 1990s.[15] The plunder of sandalwood continued unabated, with the active connivance of Forest Department staff. Substantiating this, a Central Bureau of Investigation (CBI) report says: 'It appears some of the Field Staff of Forest Department were hand in glove with notorious smugglers and their agents in Vachathi and nearby villages. In fact, investigation has revealed that there was a chit/token system in vague to ensure smooth operations of illegal felling, transport, concealment in river beds, pattah lands and nullahs for some time, before loading and ferrying the sandalwood to final destinations.'[16]

Given this scenario at the macro-level, let us see the complicity of different departments of the state government in the violation of the tribals' human rights in the name of protecting forest resources and/or the prevention of illicit felling of sandalwood in Vachathi village.

On 20 June 1992, at around 10 a.m., a group of forest officials went to the Vachathi area to enquire about illicit sandalwood felling and found two people cutting sandalwood. After a chase, they lost track of the duo near Vachathi. Continuing their search, they asked an old man, his wife and son, who were digging a well, if they had seen anyone. When they expressed ignorance of the two culprits, the forest personnel abused them of shielding the accused, and severely thrashed them.[17]

On hearing their wailings, other villagers rushed to the spot and there was a scuffle. According to the Forest Department officials, a group of women armed with chilli powder and stones and a few men possessing country-made guns launched a brutal attack. In the process, a member of the Forest Department staff fainted after reportedly being hit on the head, but was taken to the village by a group of women, who gave him food and water. Later, the villagers sent him to hospital in a bullock cart.[18]

Laying siege to the other staff, forty in number, the villagers demanded an explanation for the thrashing of the three innocents. The chief official explained that they had been sent to enquire about reports of illegal sandalwood felling in the forest. Accepting this, the villagers finally let them to go. However, the Assistant Conservator of Forests lodged a complaint with the Inspector of Police at Harur against fourteen people by name and about 300 others for the alleged attack on the forest officials. Arresting the villagers, the police invoked various sections of the Indian Penal Code (IPC) and the Indian Arms Act (IAA). All of them were remanded by a judicial magistrate at Harur on 21–23 June 1992.

Significantly, on the same day as the initial disturbance, at about 3.30–4.00 p.m., about 400–600 policemen and forest officials arrived in Vachathi and began to indiscriminately assault the villagers. Men and women were beaten up mercilessly, with the CBI report saying that 'locked houses were broken open and the village was ransacked by the combined party'.[19] Some of the men fled to the hills and to other hideouts. About 189 men and women and twenty-eight children (aged two to seven years) were arrested. Some young women were singled out and taken to the river bed by male forest and police officials, on the pretext of unearthing sandalwood, and were raped. Although about forty female constables were present, none of them escorted these young women. Then they were herded to the Harur forest ranger's office, on the way being sexually harassed. Their blouses were torn off, their breasts being prodded by hand and with sticks by the forest personnel. They even attacked some of the women's vaginas with sticks.[20]

At about 7.30 p.m., more than 200 villagers including eighteen children were taken to the forest ranger's office, being kept there until 5 p.m. the next day. Although around ninety-four women were present, no female police constables were posted on duty during this period.[21] Clearly, this was illegal detention. During the night, some young women were taken to a room, stripped naked and then raped. Moreover, men and women, including the village headman, were forced to strip each other and to beat each other with brooms. On the evening of 21 June, ninety-five men, ninety-four women and twenty-eight children were lodged at the subjail and central jail in Salem. Again in an early morning operation, on 21 June at around 6 a.m., the same combined party consisting of 267 personnel again raided the village, but no one else was arrested.[22] The harassment continued for four weeks.

Fabricated Cases

On 20 June, at around 6 p.m., a case was registered against fourteen villagers of Vachathi under various sections of the IPC and IAA. In addition to this, three STOR cases were filed. In the first case, No. 39/92, against five men and twenty-one women, it was alleged that 1326.400 kg of sandalwood worth Rs.265,280 had been seized.[23] In the other, No. 40/92, against five men and twenty-six women, it was stated that 1497.600 kg of sandalwood worth Rs.299,500 had been seized from them.[24] This was filed as if they had been detected between 5 a.m. and 5.30 a.m. on 21 June 1992 in the Thombakkal Reserve Forest. The third case was No. 96/92 against five men and twenty-nine women, and it was fabricated as if it had been detected at 6 a.m. on 21 June 1992 in the Pallipatti Reserve Forest.[25] Further, the CBI found that 'in all three STOR cases, documents were fabricated and the thumb and palm impressions of the villagers obtained under coercion'.[26] 'The accused villagers implicated in the three STOR cases were remanded to Judicial Custody on 21–6–92 and all of them were taken to Salem Jail by midnight of 21–6–1992. The remand was extended up to 27–8–92 periodically and thereafter they were released on bail in groups. In the aforesaid three cases, charge-sheets were filed against 76 women and 15 men by the Forest Department.'[27]

Damages to Property

It is expected that the state will provide protection to the people and their property, but in Vachathi properties were either damaged or plundered by none other than the state agencies. Locked houses were broken into, roofs and tiles were damaged, household materials were plundered and later sold, land documents and ration cards were stolen, electricity connections were disturbed and diesel-engine pumps were destroyed or thrown into the wells. Livestock and other property, such as diesel engines and household materials, were taken away and sold in the nearby market, finding their way eventually to the old material collection shops at Harur.[28] Similarly, valuables such as gold, silver, money, an oil engine, two motorbikes and a number of bicycles were taken away.[29] All these unlawful activities were possible only because of connivance between the revenue, police and Forest Department officials. The personnel of these three departments, involved in the raids on 21 June, were said to be enraged and in a mood to take revenge.[30] After attacking the villagers,

> their common object broke down the locked houses, destroyed household articles including vessels, utensils, radio and bicycles in the village. They also broke the doors of the houses and destroyed the tiles on the roofs. They pulled down and destroyed about 50 houses. They also damaged oil engines and pump sets by pushing them into the wells and thereby rendered the water unfit for human consumption. The accused members of the unlawful assembly thus committed mischief causing a loss of about Rupees Two Lakhs to the villagers and properties.[31]

On 21 June, at about 6 a.m. the same combined of 267 again raided the village, several houses being destroyed, and in the evening the party raided the village and damaged property.[32] The PUCL fact-finding report (PUCL 1992) and the Bhamathi report have graphic details about the loss to property and livelihood.[33]

Rape of Young Women

After the raid in the evening of 20 June, at about 6.45 p.m., as previously mentioned, some forest officials took eighteen young girls towards the lake, located about half a kilometre from the village, and raped them. None of the women police constables or old women of the village were allowed to accompany them.[34] This is a chilling indictment of the police and forest personnel. Thus far we have a detailed account of how the state institutions themselves were culpable in crimes against the subalterns. In the following section, let us see how their voices were either ignored or neglected.

Representations and Response of the State

Representations were made by Vachathi villagers both collectively and individually, as they were by political parties and civil society organisations. Local politicians from All India Anna Dravida Kazhagam (AIADMK), incidentally the ruling party, and the Communist Party of India (Marxist) (CPI-M) submitted petitions to the respective authorities, seeking suitable action against the errant officials. The CPI-M took up the issue, and its Rajya Sabha member A. Nallasivan was instrumental in gaining a CBI probe into the incident. Initially, representations were made to the District Collector, District Forest Officer, State Conservator of Forests, Minister for Forests, Chief Minister and Chief Justice of Madras High Court regarding the atrocities committed by the forest, police and revenue department personnel at Vachathi between 26 June and 18 July 1992. In a telegram sent to the District Collector days after the incident, the Dharmapuri District Secretary of the Tamil Nadu Tribal Association (TNTA) brought to his notice the arrest of women from Vachathi and its neighbourhood by the Forest Department authorities and the damage to livelihood and property. Informing the District Collector about people fleeing the village in fear, the District Secretary requested him to take appropriate action.[35] The CPI-M led TNTA, at its conference held in Chitteri on 7 July 1992, passed a resolution seeking justice for the affected people. As a further measure, it organised a procession at Harur and a *hartal* on 13 July. The next day, a delegation consisting of Non-Government Organisation (NGO) representatives and TNTA secretary P. Shanmugam and M. Annamali, a former Member of the Legislative Assembly (MLA), visited Vachathi village and also met the tribal women at the Salem subjail, hearing of the rapes and other atrocities. Later, they submitted a petition to the Collector, Dharmapuri district, and sought a judicial enquiry about sandalwood smuggling.[36] On 18 July, A. Nallasivan, Member of Parliament (MP), Rajya Sabha and P. Shanmugam gave separate representations to the Tamil Nadu Chief Minister, seeking justice and relief for the tribals.

On 28 July, the villagers collectively wrote a letter to the Chief Justice of Madras High Court seeking justice and relief. On 29 July, they wrote individual letters pressing for the same. The letters from the women narrated the rape, assault and different forms of cruelty done by the forest and police department officials. On 30 July, P. Shanmugam of TNTA presented a petition to the Public Interest Litigation Cell of Madras High Court. On 5 August 1992, A. Nallasivan and other MPs raised the issue in the Rajya Sabha. On 22 August 1992, eighteen affected women lodged a complaint at Harur police station against the police and Forest Department officials, charging them with rape and other offences. But the Deputy Superintendent of Police advised and directed them to seek remedy through an executive magistrate.[37]

The state government, instead of taking appropriate action against those who had violated the basic human rights and dignity of the tribals, came to the defence of the accused. It was unfortunate that the Minister of Forests supported the barbaric approach of the law enforcement machinery. A ruling party heavyweight and the Transport Minister in the AIADMK government argued that the forest and police officials did not rape the tribal women or carry out the raids, and also dismissed the allegations of rape and raids made by the parties of the left as baseless.[38]

Silence of State Administration

Officials of the revenue, police and Forest Department, right from the District Collector to the Superintendent of Police, are either directly or indirectly associated with this incident. Several petitions and representations were made to the various government officials: Thashildar, Harur; Collector, Dharmapuri District; Forest Ranger, Harur; District Forest Officer, Dharmapuri District; Chief Conservator of Forest, Chennai; Sub-Inspector, Harur; Superintendent of Police, Dharmapuri; Chief Secretary, Government of Tamil Nadu; and Chief Minister, Government of Tamil Nadu by the villagers and members of various political parties during June and July 1992.[39] It was not surprising that scant regard was paid to these representations. Instead of taking any action, the officials were bent upon covering up the incidents.

The first official response from the state government was a statement made by the Minister of Forests on 21 July 1992, in which he supported and defended the government officials, saying that they had acted in order to curb the villagers' smuggling activities. He further charged that the parties of the left, especially the CPI-M, were raking up the issue for political mileage. The Collector was also of the opinion that the 'incident has been exaggerated by some political parties'.[40] Further he claimed that 'there was no rape case and they had sent women police for the raid. He felt that the allegations that houses were damaged may not be done, but only locked houses were opened to trace illegal arms and to find evidence about the smuggling of sandalwood.'[41]

The Collector visited the village only after a specific direction was given by the Madras High Court on 4 November 1992, and he submitted a report about the prevailing conditions on 12 December 1992. As per his instruction, new family cards were issued to those who had none and duplicate cards were issued to those who had lost them.[42] Twenty-eight persons who had lost their *pattah* (land deeds) were duly provided with copies. Until the court intervention, neither the district administration nor the state government had taken any action to redress the grievances of these subalterns.

Role of Left Political Parties, Civil Societies and Judicial Intervention

Petitions by the affected people individually and collectively were made by the villagers, village level politicians, tribal association representatives, MPs (Rajya Sabha) to the state government administrative functionaries: the Revenue Divisional Officer, District Collector, Ministry of Forests, Chief Minister and state judiciaries, with a petition to the Public Interest Litigation Cell of Madras High Court, Chief Justice of Madras High Court. The issue was also raised in Rajya Sabha. Despite this knocking at several doors, justice remained elusive as the voices of the subalterns were not heard at any level of the state administration. Finally, a Marxist MP, A. Nallasivan, approached the Supreme Court of India and filed a writ petition on 3 September 1992 under Article 32 of the Constitution of India, seeking justice for the Vachathi people and pleading for the following: a direction for an enquiry into the events and happenings commencing on 20 June 1992 and the consequent suffering undergone by the villagers; appropriate penal action against the delinquent personnel and compensation for the victims; a probe into the wider question of sandalwood smuggling in this area to discover and eliminate the nexus that existed between the culprits, politicians and officials of the administration; and by way of an interim measure and an ex-parte order ordering the respondents forthwith to resume bus services to and from the village, restore electricity, take measures to provide drinking water and basic sanitation conditions, provide basic rations and advance ad hoc amounts for repair of the damage done to their homes.[43] On 8 September 1992, the Apex Court transferred the case to the Madras High Court to be dealt with as a petition under Article 226 of the Constitution, and allowed the petitioner the liberty to make amendments. Further, the Supreme Court requested the Madras High Court to deal with the petition as early as possible.[44]

Consequently, Nallasivan filed a Writ Petition (No.14949 of 1992) at Madras High Court on 20 September.[45] On 4 November the court directed the Collector to 'visit the village and see the problems of the local villagers with regard to the basic necessities like water, drinking water, electricity and supply of essential commodities through ration cards, are properly attended to expeditiously'.[46] Further, the court appointed Bhamathi, Director of Scheduled Castes and Scheduled Tribes Commission, Madras, as Commissioner to submit a report about the prevailing conditions in the village and the problems of the local villagers.[47] In her report, she called for a comprehensive probe into all aspects arising from the raid and the charges of harassment of young tribal women in the forest ranger's office on 20 June 1992 so that the guilty were brought to book.[48] Admitting the involvement of some Vachathi villagers in sandalwood smuggling, the report maintained that they were only a conduit, exploited by the powerful masterminds engaged in illegal operations. Unless the master operators were brought to book, there was no point in punishing the villagers alone for the illegal trade, the report concluded.[49] On 24 February 1995, Madras High Court ordered an enquiry by the CBI.

CBI Enquiry

The court directed the CBI to investigate the complaint of eighteen women of Vachathi village to the sub-inspector of police of Harur police station alleging torture, rape, illegal detention, destruction of property and so on in connection with the combined raid of the Forest Department, police and revenue officials, and the detention of ninety women and twenty-eight children of Vachathi village in the forest ranger's office, Harur, on 21 June 1992. Further, the court directed the CBI to discover who had been involved in these crimes and to place them before the appropriate judicial authority for trial. The premier investigating agency took two months to finish this task.

Completing the enquiry, the CBI submitted its final report to the Madras High Court along with a copy of the charge-sheet dated 23 April 1996 that had been filed before the Chief Judicial Magistrate. Interestingly, 269 accused belonging to the forest, police and revenue departments were charged under various sections of the IPC and the Scheduled Castes and Scheduled Tribes (Prevention of Atrocities) Act 1989.

Interim Compensation

Based on the CBI report, on 3 May 1996 the Madras High Court ordered the state government to give each of the eighteen victims Rs.10,000 as an interim compensation within two months. Further, the state government was directed to take appropriate action against the 269 forest, revenue and police officials whose names appeared in the charge-sheet. The court also suggested the appointment of a commission of enquiry to discover the truth and extent of the connections between sandalwood smugglers, politicians and officials. The above narratives clearly brought out how the state institutions had been ineffective as far as subaltern issues were concerned, with the victims consequently having to struggle to get justice.

Subalterns Through the Prism of Power

Given the nature of the atrocities visited on the Vachathi tribals, how the state administration fabricated the incident and made efforts to cover up the nexus between the forest, police and revenue departments and ruling party politicians needs to be substantiated and explained in detail.

Forest Department

M. Ramamurthy, forest ranger, Uttangarai, in his remand report to the Judge (Criminal), Harur, dated 21 June 1992, stated that the forest staff and special squads had caught twenty-eight culprits, both men and women, while carrying sandalwood by head load from the reserve forest and hiding on the river bed early in the morning, at around 5 a.m. The seizure weighing 1326.400 kilograms was worth Rs.265,280. According to the Tamil Nadu Forest Act 1882 Section 21(d,e,f); Forest Act (Amendment) 1973, 1970, 3(1,7) and 1979 these are offences.[50] Since the possible identities and localities of the culprits were fake, he prayed the judge to keep them under remand for fifteen days. In yet another remand report of the same date, C. Guru Anaimuthu, forest ranger, stated that thirty-one persons, including men and women, were caught while carrying the sandalwood and hid it in the bushes at around 5.30 a.m. The total weight of the sandalwood was about 1497.600 kilograms, worth about Rs.299,500. The Forest Department clearly fabricated the document as if they had not violated any normal procedures or the provisions of the existing Acts.

Revenue Department

According to the Collector's direction, a team of revenue officials accompanied the forest and police squads to raid the village. In case of a law and order issue, the revenue officials were authorised to issue shooting orders. The Revenue Divisional Officer (RDO), Dharmapuri, N. Theivasikamani, maintained that the villagers did not retaliate as most of the men had absconded. He further admitted that he was unaware of the rampage by the police and forest officials.[51] The RDO's report, dated 10 August 1992, clearly established the existing nexus between the forest, police and revenue departments. It denied instances of manhandling, damage to property and rape. According to the report, none of the residents were found in the village, except aged persons, and all the houses were locked; but, later on it concludes that ten men and forty-seven women were remanded by the Forest Department staff. The report stated further that smuggled sandalwood was unearthed from the river bed and other *pattah* lands, about 17,200 kilograms of wood being seized and taken to the forest ranger's office on four lorries.

During the second raid on the same day, the report stated that 38,400 kilograms of sandalwood was seized, with the arrest of eighteen men and forty-six women. Further raids were conducted on 30 June, 1 July, 2 July and 7 July 1992, yielding 3400 kilograms of sandalwood and the arrest of twenty-seven men and one woman.

Regarding damages, the RDO's report reported that only forty-nine houses were ransacked, but the specious argument was given that the tribals themselves did the damage in order to put the blame on the Forest Department.[52]

In the same vein, the grave allegations of the CPI-M against the forest and police departments were dismissed as a smokescreen to hide illicit smuggling. The RDO argued that the Vachathi village tribals, particularly the men, invariably entered the forest well before dawn to cut down sandalwood and returned only after sunset. Hence, tribals were not found in the daytime, he contended, going on to charge that the CPI-M actively encouraged smuggling. In fact, the party flag had been hoisted in the area very recently, having been almost completely absent previously. In the RDO's view, the CPI-M was raising allegations against the forest and police departments with a vested interest.

Describing the allegations of rape as fabricated, he maintained that they were raised only fifteen days after the incident. Moreover, the allegations initially came from three women, then from fifteen and later on from twenty-three; it was also contended that no raid took place during the night. Even during the raid, fifteen women police constables were present. That there was no complaint from the neighbouring villages was cited in order to whitewash the whole incident. While the District Collector maintained that he was not aware of the incident, the RDO, who was involved in the operation, clearly took sides and made an attempt to absolve the official machinery of any wrongdoing.

District Forest Officer

The District Forest Officer argued that the Vachathi 'villagers were no more interested in agriculture as they had a taste of "big money" through sandalwood smuggling'. He further made a derogatory comment about the tribal women saying, '"these women" were of "loose morals" and would "do anything" and that they even kicked the forest officials'.[53]

State Forest Department

R. Baskaran, Deputy Secretary to the Government, Environment and Forest Department, argued that 'the villagers of Vachathi who are particularly notorious in carrying out illicit felling and storing illicitly removed sandalwood. They are also said to be aiding smuggling of sandalwood to Madras, Kerala and Karnataka, whenever the situation was congenial for such illegal transport.'[54] He further contended that the raid party was engaged in collecting the buried sandalwood when the villagers, numbering about 300, came fully armed with deadly weapons such as guns, knives, spears and sticks, and started to attack the forest staff, without any provocation.[55] 'It is denied by these respondents that the people of Vachathi had been victims of brutal attack by the Forest and Police Personnel.'[56] He went on to dismiss the alleged rape and ill-treatment of the Vachathi villagers as fabricated and false. 'Neither during the time of arrest of women accused nor at the time of remand before the Judicial Magistrate they have complained about sexual assaults or bodily attack',[57] he said, disagreeing that the women were illegally detained in the forest office, wherein some of them were stripped, molested and raped. In his view, these allegations were imaginary, false and baseless and made to garner cheap publicity for politicians with vested interests.[58] While admitting that smuggling was going on unchecked in Chitteri Range with the connivance of the Forest Department, police officials and powerful politicians, he maintained that the petitioner had chosen to defend the actual smugglers while shifting the blame to the police, forest officials and other politicians.[59]

Minister of Forests

K.A. Senkottaiyan, Minister of Forest, Government of Tamil Nadu, said that smuggling was a daily routine for all 300 families in Vachathi. It had been going on for the last five years and everyone earned about Rs.500 per day from it. Damages to the villagers' property and their ill-treatment by the forest, police and revenue department's people were denied as untrue. For him the incident of mass rape was a figment of their imagination, and he was of the view that during a raid that was intended to ascertain sandalwood smuggling the village residents attacked the officials. The minister also maintained that the allegations were made in order to discredit government officials.[60]

There are glaring discrepancies in the remand reports of Uttangarai forest ranger Ramamurthy and Harur forest ranger Guru Anaimuthu. According to the former, during the raid twenty-six people were caught and the seizure was 1326.4 kilograms, worth about Rs.265,280. But the latter stated that thirty-one villagers were arrested and 1497.6 kilograms of sandalwood worth Rs.299,500 were recovered.[61] Giving yet another contradictory account, the RDO report says that 38,400 + 38,400 + 3400 kilograms of sandalwood were seized. The factual errors apart, all these reports had one thing in common: the fabrication of a lie and then sticking to it in order to protect the officers and personnel who had been involved in grave human rights violations.

Forest Laws and Tribals

Historically, the forest resources were used and managed by the local communities without any restrictions. Stressing this, Shiva says the 'forest had been controlled and used collectively by village communities thus ensuring a sustainable use of these renewable resources'.[62] Of course there had been several conflicts over forest use between different communities/groups in various parts of the country.[63] However, no ruler had imposed any curbs during the pre-colonial period except on some specific trees in some regions. For example, Tipu Sultan had exercised control over sandalwood trees, realising their commercial value during the late eighteenth century.

After colonial intervention, the government claimed that the forests belonged to the state. Interestingly, conservation was not the intention behind these claims; rather they were cover for commercial interests and greed, so that forest wealth could be legitimately looted. To bring the forests under their control, they accused the tribals and other forest users of causing destruction to the green cover, which had to be prevented. Huge quantities of forest resources, especially sandalwood, were exported to other countries from the early nineteenth century until the mid-twentieth century.[64] Besides sandalwood, trees in vast tracts were cleared for the British ironmaking industries during the second quarter of the nineteenth century.[65] A large amount of forest cover was brought under coffee and tea plantations during the third quarter of the nineteenth century.[66] In addition, the axe fell on several thousand trees to allow the laying of railway lines in the same period.[67] However, there was no restriction on the tribals except where plantations were established. In other words, although the commercialisation of forest was encouraged until 1865, when the first Indian Forest Act was enacted, the tribals' access to the forest was not curtailed.

Under the Forest Act of 1882, the tribals could only claim the right to pathway, watercourses, pasture and forest produce (Section 10).[68] Further, these rights were intermittently granted, with quantitative restrictions (Section 12). Fresh clearings for cultivation or any other purpose were also restricted (Section 7). If the tribals used the prohibited items, they were liable for imprisonment for six months or a fine of Rs.500, or both (Section 21). Grazing on forest land was limited and liable for suspension (Section 22). Grazing cattle on closed land would attract imprisonment for a month and/or a Rs.200 penalty (Section 28). According to Section 28A(1), in any case where such infringement related to any scheduled timber, imprisonment might extend to five years and a fine might be up to Rs.20,000. For a first offence, the term of imprisonment should not be less than two years and the fine should not be less than Rs.7500; while for a second or subsequent offence, the term of imprisonment should not be less than three years and the fine should not be less than Rs.15,000. The tribals were forced to render free services to the police as well as Forest Department officers. Under Section 36A,

> No person shall have in his possession any quantity of sandalwood in excess of five kilograms, unless under a license granted by the District Forest Officer in that behalf or unless such sandalwood is affixed by a Forest Officer with such mark and in such manner as may be prescribed.

The Madras Forest Act 1882, Section 41(1) stated:

> When there is reason to believe that a forest offence has been committed in respect of any timber or forest produce, such timber or produce, together with all tools, ropes, chains, beats, vehicles and cattle used in committing any such offence, may be seized by a Forest Officer or Police Officer.

Section 41(2) stated:

> Any Forest Officer or Police Officer may, if he has reason to believe that a vehicle has been or is being used for the transport of any scheduled timber in respect of which there is reason to believe that a forest offence has been or is being committed, require the driver or other person in charge of such vehicle to stop the vehicle and cause it remain stationery as long as may reasonably be necessary for examination of the contents in the vehicle and inspection of all records relating to the scheduled timber and in possession of such driver or other person in charge of the vehicle or any other person in the vehicle.

Under Section 51:

> Any Forest Officer or Police Officer may, without orders from a Magistrate and without a warrant, arrest any person reasonably suspected of having been concerned in any forest offence punishable with imprisonment for one month or upwards if such person refuses to give his name and residence or gives a name or residence which there is reason to believe to be false, or if there is reason to believe he will abscond. Any person arrested under this section shall be informed as soon may be of the grounds for such arrest and shall be produced before the nearest Magistrate within a period of 24 hours of such arrest excluding the time necessary for the journey from the place of arrest to the court of the Magistrate; and no such person shall be detained in custody beyond the said period without the authority of a Magistrate.

Under Section 52:

> Any Forest Officer or Police Officer who vexatiously and unnecessarily seizes any property on pretence of seizing property liable to confiscation under this Act, or who vexatiously and unnecessarily arrests any person shall be punished with imprisonment for a term which may extend to six months or with fine which may extend to five hundred rupees.

The post-independence period witnessed a progressive curtailment of tribals' right over the forests. Even the use of timber for housing and to make agricultural implements was proscribed under the Tamil Nadu Hill Areas (Preservation of Trees) Act 1955. These restrictions were applicable even to the steep slopes of the hills (Section 4(1). Anyone who offended against Section 3 or 4 of this act was liable for punishment with imprisonment up to one year or penalty up to Rs.5000 or both (Section 7). After independence, tribal rights to cut down the trees even for basic necessities were further curtailed.

The Madras Forest (Amendment) Act 1979 increased the penalty for the offences committed under the Madras Forest Act 1882. For example, the penalty for using prohibited forest items was raised from six months to three years' imprisonment and from Rs.500 to Rs.10,000, or both. Therefore, the rights of tribals regarding forests during the post-colonial period were curtailed by increasing penalties. This resulted in the systematic destruction of the forest-oriented tribal economy.

The government enacted several rules and regulations for the transit of timber from one region to another. According to the Tamil Nadu Timber Transit Rules 1968, 'No person shall move timber into or from, or within the State by land, water or air unless such timber is accompanied by a permit prescribed there for under rule 4' (Section 2). Further, the permit for the import as well as export of timber was to be produced at any checking station if notified by the Collector (Section 5[i]): 'In the case of timber in transit, the permit shall be produced instantaneously when demanded, anywhere, within the limits of the scheduled areas, by any Forest Officer or Police Officer or at any of the checking stations notified by the Collector outside the schedule areas past which timber is taken' (Section 5[i]).

Sandalwood Protection

In addition to the general Acts, the Tamil Nadu government made specific rules, the Sandalwood Transit Rules 1967,[69] and Sandalwood Possession Rules 1970,[70] to protect sandalwood resources. These rules were framed to exercise the powers conferred by Sections 35, 36, 36A, 36B and 36D of the 1882 Forest Act, to prevent sandalwood smuggling. According to Section 9(a) of the Sandalwood Transit Rules 1967, not exceeding 5 kilograms of sandalwood can be carried by any bona fide traveller or under his authority for his personal use. 'The District Forest Officer shall maintain an approved list of companies or persons engaged in the distillation of sandalwood or in the manufacture of any of its bye-products. He may at his discretion include the name of any company or person in the said approved list, after giving reasonable opportunity to such company or person' (Section 10). 'No person shall import sandalwood into or export sandalwood from, or move sandalwood within any place in the State unless such sandalwood is accompanied either by a permit referred to in Rule 4 or by a way permit referred to in rule 6 and unless such sandalwood bears the mark of the Government or origin, as provided in rule 7 of the property mark in respect of such sandalwood which has been duly registered in accordance with rule 7. Such sandalwood shall also be produced for examination at its destination at the appropriate checking station specified in Schedule A' (Section 3). 'In the case of sandalwood purchased at the Government sales depot in the State of Andhra Pradesh, Mysore or Kerala and imported into the State, the permit shall be that issued by the Government of the State of Andhra Pradesh, Mysore or Kerala or any authority in that State responsible for the sale of sandalwood in that State and such permit shall be valid only for transport of sandalwood to the destination specified therein' (Section 4[2]).

If the sandalwood is imported other than in Section 4(2), a permit form has to be collected from the district forest officer on payment of a price fixed by the chief conservator of forest (Section 4(3). For the subsequent movement of the imported sandalwood and the transport of all local sandalwood, a permit is to be obtained from the district forest officer on payment (Section 4[4]). Sandalwood transport is also restricted to the daytime until it gets clearance from the first checking station. 'All Sandalwood in transit shall be produced for examination at the first checking station in the district between the hours of 6 am and 6 pm and if such sandalwood arrives at the first checking station after 6 pm the officer in charge of such checking station may detain it till next day for examination' (Section 6[1]).

If the officer of the checking station concerned is not satisfied, he can detain the sandalwood and inform the District Forest Officer (Section 6(2). Once the sandalwood has been checked at the checking station, it has to be endorsed at the subsequent checking stations (Section 6[5]). Different department officers can check the sandalwood transport. 'Any Revenue Officer not below the rank of Revenue Inspector in charge of firka or any Police Officer in uniform not below the rank of Head Constable or any Forest Officer not below the rank of Forest Guard 1 or Tanadar in charge of a checking station mentioned in Schedule A, may stop any vehicle reasonably suspected of carrying sandalwood for checking its contents and the permits referred to in these rules shall be produced when demanded and such vehicle shall not proceed there from until permitted to do so by the said officer of the Revenue, Police or Forest Department or the said Tanadar' (Section 6[7]).

The transport of sandalwood involves passing through the checking station. 'No person shall take any sandalwood off or pass it from any road on which a checking station has been placed with intent to evade the production of the sandalwood at a checking station' (Section 6[8]). According to Tamil Nadu Sandalwood Possession Rules 1970 Section (1) no person shall possess stock of sandalwood in excess of 5 kilograms without a licence. In short, the Forest Department intended to protect forest resources in general and sandalwood in particular, and in Tamil Nadu several Acts and Rules were enacted to fulfil this.

Given the presence of these stringent rules and regulations, the extraction and transport of forest resources is not possible unless there is a clear understanding between the smugglers and the police, forest and revenue departments. Even then it could not be carried out in the absence of the tacit support of the political leadership.

Constitutional Rights, Special Acts and Commissions for the Tribals

Several safeguards are guaranteed to the Scheduled Castes and Scheduled Tribes by the Constitution of India. Article 46 says: 'The State shall promote with special care the educational and economic interests of the weaker sections of the people, and in particular, of the Scheduled Castes and the Scheduled Tribes, and shall protect them from social injustice and all forms of exploitation.' Based on this, the government has enacted several legislations: the Protection of Civil Rights Act 1955, Scheduled Castes and Scheduled Tribes (Prevention of Atrocities) Act 1989, Scheduled Castes and Scheduled Tribes (Prevention of Atrocities) Rules 1995. A statutory panel, the National Commission on Scheduled Castes and Scheduled Tribes, to protect these groups, who are placed under the same terms given their socio-economic conditions.

Scheduled Castes and Scheduled Tribes (Prevention of Atrocities) Act 1989

In 1989, the central government enacted the Scheduled Castes and Scheduled Tribes (Prevention of Atrocities) Act 1989. Besides protecting the marginalised tribals from economic exploitation, the legislation was aimed at ensuring their human dignity. For example, the following were listed as grave offences under the Act: disrobing or parading naked or any similar act which is derogatory to human dignity; wrongful dispossession of land or premises or interfering with the enjoyment of rights over any land, premises or water; compelling or enticing to render forced labour or bonded labour; filing false cases; giving false information to a public servant causing the misuse of lawful power to the detriment of a member of Scheduled Castes/Scheduled Tribes; intentional insults or intimidation with the intent of public humiliation; assaulting or forcing a woman to dishonour or outrage her modesty; using one's dominant position to sexually exploit Scheduled Castes/Scheduled Tribes women; corrupting or spoiling drinking water sources—any spring, reservoir, tank or well; and forcible eviction from houses. These would all attract imprisonment from six months to five years.

Section 2 of the Act even provides for the death penalty to anyone who tenders false evidence leading to the conviction and resultant execution of an innocent tribal in a fabricated case. Wilful negligence of duties required under this Act by a non-Scheduled Tribes public servant would attract imprisonment for one year. Further, Section 21(2) and (3) deal with the provision of economic and social rehabilitation for the victims of atrocities.

National Commission on Scheduled Castes and Scheduled Tribes

This statutory panel is empowered to direct state law enforcement agencies to furnish relevant details and to indicate the action taken whenever an atrocity is committed. It has to ensure whether the Collector and Superintendent of Police of the district concerned have visited the spot on receipt of information; proper filing of a first information report in the local police station with inclusion of all the accused; investigation by a senior police officer as per provisions of the Scheduled Castes and Scheduled Tribes (Prevention of Atrocities) Act 1989; apprehension of culprits without loss of time; filing of a proper charge sheet invoking relevant sections of the IPC together with the Protection of Civil Rights Act 1955 and Scheduled Castes and Scheduled Tribes (Prevention of Atrocities) Act 1989 in court; appointment of special public prosecutors to verify whether the cases are tried by special courts; and monitoring of the police assisting the courts in bringing forward witnesses and seeing that the culprits are suitably punished.

Further, the Scheduled Castes and Scheduled Tribes (Prevention of Atrocities) Rules 1995 was framed to give additional special protection to the tribals. Despite the constitutional safeguards and the existing legal mechanism, implementation rests with the lower-level bureaucracy and law enforcement wings, which themselves stand accused of observing the statutes more in the breach than the observance. Interestingly, the Director of the Scheduled Castes and Scheduled Tribes Commission, Madras, visited Vacathi, assessed the situation and submitted a report to government within one and half months.

State Institutions–Politicians Nexus and Violation of Human Rights

Representations from Vachathi villagers to the District Forest Officer, District Collector, Superintendent of Police and Minister of Forests about the prevalence of illicit felling in the area since late 1991 fell on deaf ears.[71] To press for action against rampant smuggling, the people organised a protest fast early in December 1991. Several Tamil dailies which reported sandalwood smuggling pointed out that the culprits were either local political heavyweights or their relatives. Prominent among them were members of the ruling AIADMK who were related to politicians, particularly to the ruling party organ of AIADMK, the joint secretary of Namakkal town AIADMK unit, the younger brother of the former AIADMK-MLA, a former Salem district secretary of AIADMK, a former AIADMK-MLA and a son of former Tamil Nadu government's special representative in New Delhi. It has been established that the smugglers have made crores (one crore equal to 10 million) of rupees out of the illegal operations. Even when arrested they were granted bail immediately.[72] The forest officials also admitted that some of them acted hand in glove, and reportedly received Rs.50,000 per truckload of wood as a bribe.[73] The Forest Department failed to act on the protests made by Vachathi residents and media reports. In fact, it appears the smugglers not only enjoyed the political patronage of the ruling party but also to some extent by judiciary as well. The leniency shown by the courts to those accused of sandalwood smuggling indeed raises certain questions.

Given the nature of Forest Acts and STOR rules/sandalwood rules, it is very difficult for illicit transit of any kind of forest resources, sandalwood in particular, to take place. It is possible to bypass a checkpoint or two by taking a diversion in the forest but not all of them. Destinations for the smuggled consignments are always neighbouring Kerala, Karnataka, Andhra Pradesh, Pondicherry and Goa, where a sizeable number of sandalwood manufacturing units/industries are located. For instance, Kerala has twenty-four units while Goa has seven, Andhra Pradesh five and Pondicherry four.[74] Someone smuggling sandalwood to Kerala has to pass through at least three districts, Salem, Erode and Coimbatore, and there are several checkpoints en route. Transporting smuggled sandalwood to Karnataka and Andhra Pradesh would require passing at least two checkpoints besides the checking carried out on inter-state borders. 'With the objective of effectively protecting the forest wealth, particularly sandalwood and timber, seven forest protection squads besides a mobile patrol team were in operation till 1977. Later on, the numbers of the squads were increased to twelve and three detective flying squads were formed.'[75] The numbers of checkpoints were increased to eighty-nine in 1986–1987.[76] According to the State Transport Authority, Government of Tamil Nadu, the Transport Department maintains seventeen checkpoints on the state borders in Tamil Nadu, of which sixteen are on the borders with Andhra Pradesh, Karnataka and Kerala,[77] and one is close to the Union territory of Pondicherry.[78] In addition to this, there are twenty-five Commercial Tax Officers (CTO) and Assistant Commercial Tax Officers (ACTO), working at different checkpoints at the borders of Tamil Nadu—eight in Andhra Pradesh, nine in Kerala, two in Karnataka and six in Pondicherry.[79]

From this it becomes crystal clear that smuggling could never be carried out without an understanding within the Forest Department. If not the Forest Department, then who is facilitating the illegal activities? Obviously it is the politicians. Even a layman would observe this without much difficulty.

While ignoring the plea of the tribals to curb the illicit felling of sandalwood, the state administration—right from the police stations and revenue and forest divisions to the Deputy Secretary to Forest Department and Ministry of Forests—has remained content to point the finger at the voiceless subalterns and to stifle them. Moreover, the tribals have had to suffer the brutality of the state apparatus.

The number of cases registered under the Protection of Civil Rights Act 1955 and the Scheduled Castes and Scheduled Tribes (Prevention of Atrocities) Act 1989 shows a progressive decline. However, one should not construe from this that the level of atrocities has decreased. In Tamil Nadu, eighty-eight cases were registered in 1997 under the Protection of Civil Rights Act 1955, and this declined to fifty in 1998, then to forty-five in 1999 and thirty-two in 2000. Under the Scheduled Castes and Scheduled Tribes (Prevention of Atrocities) Act, 1989, 643 cases were registered in 1997, decreasing to 574 in 1999. Yet the high number of cases still registered under these Acts is a grim reminder of the unfortunate fact that atrocities against the Scheduled Castes and

Scheduled Tribes continue unabated even today. Though the number of cases of violation of human rights against tribals has decreased over the years, state-sponsored atrocities continue in one or other form, with Vachathi being just one example.

Conclusion

It is a well-established fact that the tribals' livelihood is associated with forests and forest resources. Tribal development policies, which gave an impetus to infrastructural facilities—roads, communications, schools, hospitals, credit and so on—made an attempt to integrate the tribal communities' economy into that of mainstream society. The TSP of the last two decades of the twentieth century in particular was aimed at accomplishing this agenda. However, it was ironic that these measures, which were intended to spur on economic development, have somehow failed in many parts of the country.[80] Given the situation, the various arms of the state apparatus have not only failed to deliver prompt services to improve living standards but have also exploited tribal members economically and physically. The Vachathi incident is one of the worst state-sponsored atrocities on the tribal people during the post-independence period. While failing to arrest the illegal extraction of the forest resources by the organised timber mafia, which has political support, the law enforcement state institutions have exercised atrocities on the subalterns, with the effect that they have rebelled against the state apparatus. In other words, the approach of the state apparatus towards subalterns is forcing them to rebel against those institutions.

Notes

1. Counter affidavit filed on behalf of the respondents by the deputy secretary to government, Environment and Forest Department in W.P.No.14949 of 1992 (Chennai: Government of Tamil Nadu, 12 January 1993), p. 3.
2. Report of CBI on the Investigation in case No.RC.3/5/95 of CBI, SCB, Madras in Compliance with the orders passed by the Hon'ble High Court of Madras in Writ Petition No.14949/92 Filed by A. Nallasivam, Member of Parliament (Madras: Superintendent of Police, 23 April 1996), p. 4.
3. Velayutham Saravanan, 'Thamizhaga Malaival Makkalin Nila Urimai', *Uzhavan Urimai* (Land Rights of Tribals in Tamil Nadu)[Special Issue], 1999, pp. 53–57 (in Tamil); Velayutham Saravanan, 'Commercial Crops,

Alienation of Common Property Resources and Change in Tribal Economy in the Shervaroy hills of Madras Presidency during the colonial period', *Review of Development and Change*, 4(2), 1999, pp. 298–317; Velayutham Saravanan, 'Tribal Revolts in India with reference to Salem and Baramahal Districts of Madras Presidency during the late 18th Century', *Artha Vijnana*, XLI(1), 1999, pp. 67–81; Velayutham Saravanan, 'Commercialisation of Forest, Environmental Negligence and Alienation of Tribal Rights in Madras Presidency: 1792–1882,' *Indian Economic and Social History Review*, 35(2), 1998, pp. 125–146; Velayutham Saravanan, 'Colonialism and Coffee Plantations: Decline of Environment and Tribals in Madras Presidency during the Nineteenth Century', *Indian Economic and Social History Review*, 41(4), 2004, pp. 465–488; Velayutham Saravanan, 'Colonial Commercial Forest Policy and the Tribal Private Forests in Madras Presidency: 1792–1881', *Indian Economic and Social History Review*, 40(4), 2003, pp. 403–423.

4. Velayutham Saravanan, 'Tribal Land Alienation in Madras Presidency during the Colonial period: 1792–1947,' *Review of Development and Change*, VI(1), 2000, pp. 73–104.
5. Board of Revenue, Vol. 1587 (Chennai: Tamil Nadu State Archives, 11 December 1837). pp. 15047–15056.
6. Velayutham Saravanan, 'Economic Exploitation of Forest Resources in South India during Pre-Forest Act Colonial Era, 1793–1882,' *International Forestry Review: The International Journal of Forest Science and Policy*, 10(1), 2008, pp. 65–73; Saravanan, 'Commercialisation of Forest', pp. 125–146.
7. Velayutham Saravanan, 'Environmental History of Tamil Nadu State, Law and decline of Forest and Tribals, 1950–2000', *Modern Asian Studies*, 41(4), 2007, pp. 723–767.
8. Collector *Report on Vachathi Village* (Dharmapuri: 12 November 1992). p. 3.
9. Ibid.
10. People's Union for Civil Liberties (PUCL), Report of the Fact-finding team that enquired into Police/Forest officials atrocities on Vachathi Village, Harur Taluka, Dharmapuri District (PUCL, Tamil Nadu Unit, June 1992), p. 3.
11. Velayutham Saravanan, 'Colonialism and Conservation: Commercialisation of Forests and decline of tribals in Madras Presidency, 1882–1947' (paper presented at the workshop on the 'Nature in the Raj: Colonial Environments and Indigenous Knowledge' New Delhi, India, August 18–19, 2006).
12. Saravanan, 'Environmental History', pp. 723–767.
13. Velayutham Saravanan, 'Terrorising Tribals: Nexus between Police and Forest Department,' *Economic and Political Weekly*, 32(29), 1997,

pp. 1789–1790; Velayutham Saravanan, 'Tamil Nadu: An Ecological Tragedy', *Frontier*, 33(3), 2000, pp. 4–6; Velayutham Saravanan, 'Joint Forest Management in Andhra Pradesh: Who does the Offences?', *Asian Economic Review*, 48(2), 2006, pp. 231–238; Saravanan, 'Environmental History', pp. 723–767.

14. Saravanan, 'Environmental History', pp. 723–767.
15. Ibid.
16. Report of Central Bureau of Investigation, p. 2.
17. PUCL, 'Report of the Fact-finding', p. 5.
18. Ibid.
19. 'Report of Central Bureau of Investigation', p. 8.
20. PUCL, 'Report of the Fact-finding,' 5.
21. Report of Central Bureau of Investigation, p. 9.
22. Ibid.
23. M. Ramamurthy, Remand Report to the Judge (Criminal) (Harur: 21 June 1992).
24. C. Guru Anaimuthu, Remand Report to the Judge (Criminal) (Harur: 21 July1992).
25. Report of Revenue Divisional Officer to the Collector (Dharmapuri, 10 August 1992). p. 10.
26. Ibid.
27. Ibid.
28. Petition from V. Gunasekaran, Secretary, AIADMK, Vachathi to Chief Minister, Government of Tamil Nadu, 15 July 1992.
29. B. Bhamathi, Report of the on–the spot enquiry into the atrocities on Sch. Tribes of Vachathi village, Harur Taluk, Dharmapuri District by Forest officials and Police (Madras: Director for SCs and STs, 19 August 1992). p. 4.
30. Report of Central Bureau of Investigation, p. 12.
31. Central Bureau of Investigation, Charge Sheet in RC.3/5/95/CBI (Chennai: 23 April 1996). p. 5.
32. Ibid., p. 6.
33. B. Bhamathi, *Report on the prevailing conditions in Vachathi village* (Madras: Director for SCs and STs, 22 December 1992), pp. 5–6.
34. *Report of* Central Bureau of Investigation, p. 8.
35. 'Report of Revenue Divisional Officer, pp. 1–5.
36. Malai Murasu, 1 September 1992 (Tamil Newspaper), Salem edition.
37. Nallusamy D. S. P. Harur to DFO Harur, 12 March 1993.
38. Malai Murasu, 1 September 1992.
39. Dharmapuri District Tribal Association gave a telegram to the Collector, 26 July 1992; petition from the following village people and various political parties (signed by about seventy-three persons) of Vachathi, Pethanampatty, Kookkatappatty, Ko. Koootturoad, Paraiappatty Panchayat,

H. Gopinathampatty, Chinnankuppam Panchayat, Kokkarappaaty Panchayat, Malakapady, Athanoor, Chettipatty, Erumaiyampatty, Anjalwady Panchayat, Odasalpatty, T. Pudur and Kudumiyampatty village people to Chief Minister of Tamil Nadu, 6 July 1992 (copy made to Chief Conservator of Forests, Chennai; Secretary, Forest Department, Government of Tamil Nadu, K.V. Thangabalu, MP, Dharmapuri; Collector, Dharmapuri; Superintendent of Police, Dharmapuri; District Forest Officer, Dharmapuri; Deputy Superintendent of Police, Harur; Ranger, Harur; Thashildar, Harur; Sub-Inspector, Harur and Abaranji, M.L.A., Harur). Petition from V. Gunasekaran, Secretary, AIADMK, Vachathi to Chief Minister, Government of Tamil Nadu, 15 July 1992; P. Shanmugam, General Secretary, Tamil Nadu Tribal Association to the Chief Minister, 18 July 1992; Petition from the affected Vachathi Village people (signed and thumb impression consists of 220 persons) to the Collector, 20 July 1992; Ponnusamy, Secretary to Chitteri Tribal Committee, Tamil Nadu Tribal Association to Collector, Dharmapuri District, 14 July 1992; petition from affected tribal women (signed by eighteen women), Vachathi village to Sub–Inspector, Harur Police Station, dated 22 August 1992; petition from P. Shanmugam, General Secretary, Tamil Nadu Tribal Association to the Justice Badmini Jesudurai, Public Interest Litigation Cell, Chennai High Court, 30 July 1992.

40. Bhamathi, *Report of the on–the spot enquiry*, p. 9.
41. Ibid.
42. *Collector Report on Vachathi Village*, p. 3.
43. Writ Petition (Civil) No.670/92, Supreme Court of India (New Delhi: Supreme Court of India, 3 September 1992), p.19.
44. Ibid.
45. Writ Petition No. 14949 of 1992, Madras High Court (Chennai: Madras High Court, 20 September 1992).
46. Ibid.
47. Ibid.
48. Bhamathi, *Report of the on–the spot enquiry*, p. 15.
49. Ibid., 16.
50. Ramamurthy, *Remand Report.*
51. PUCL, 'Report of the Fact–finding', p. 8.
52. Report of Revenue Divisional Officer, pp. 1–5.
53. PUCL, 'Report of the Fact–finding', p. 8.
54. Counter Affidavit Filed, p. 2.
55. Ibid.
56. Ibid.
57. Ibid.
58. Ibid.

59. Ibid.
60. Dhinamani, 28 July 1992 (Tamil Newspaper), Chennai edition.
61. Anaimuthu, *Remand Report.*
62. Vandana Shiva, *Ecology and the Politics of Survival: Conflicts over Natural Resources in India*, New Delhi: Sage Publications India Pvt. Ltd., 1991, p. 14.
63. Chetan Singh, Forests, Pastoralists and Agrarian society in Mughal India in David Arnold and Ramachandra Guha (eds), *Nature, Culture, Imperialism: Essays on the Environmental History of South Asia*, Delhi: Oxford University Press, 1995, pp. 21–48; Sumit Guha, 'Forest politics and Agrarian empires: The Kandesh Bhills, c.1700–1850', *Indian Economic and Social History Review*, 33(2), 1996, pp. 133–153; Sumit Guha, 'Claims on the commons: Political Power and Natural Resources in Pre–Colonial India', *Indian Economic and Social History Review*, 39(2&3), 2002, pp. 181–196; Sumit Guha, *Environment and Ethnicity in India, 1200–1991*, Cambridge: Cambridge University Press, 1999, chapter–5; Aloka Parasher–Sen, 'Of tribes, Hunders and Barbarians: Forest dwellers in the Mauryan period', *Studies in History* (n.s.), 14(2), 1998, pp. 173–191.
64. Saravanan, 'Colonialism and Conservation'.
65. Saravanan, 'Commercialisation of Forest', pp. 125–146.
66. Saravanan, 'Colonialism and Coffee Plantations', pp. 465–488.
67. Saravanan, 'Commercialisation of Forest', pp. 125–146.
68. Forest Department, Tamil Nadu Forest Act 1882 (Chennai: Government of Tamil Nadu).
69. Forest Department, Tamil Nadu Sandalwood Transit Rules, 1967 (Chennai: Government of Tamil Nadu).
70. Forest Department, Tamil Nadu Sandalwood Possession Rules, 1970 (Chennai: Government of Tamil Nadu).
71. Petition, 26 August 1991, 27 February 1992, 15 May 1992 and 18 June 1992 (PUCL).
72. PUCL, 'Report of the Fact–finding', p. 10.
73. Ibid., p. 7.
74. http://www.indiatogether.org/2005/may/env–sandal.htm (Accessed on December 2006).
75. Saravanan, 'Environmental History', pp. 723–767.
76. Ibid.
77. Tiruthani Checkpoint and Pethikuppam Checkpoint in Chengalpet District; Tiruchittrambalam Checkpoint in Villupuram District; Ranipet (Serkadu) Checkpoint and Katpadi Checkpoint in North Arcot District; Hosur Checkpoint (Zuzu Vadi) Checkpoint and Bagalur Checkpoint in Dharmapuri District; K. G. Chavadi Checkpoint, Gopalapuram Checkpoint and Ambarampalayam Checkpoint in Coimbatore District; Bannari Checkpoint in Periyar District; Calicut Main Road (Gudalur) Checkpoint

and Talaguntha Checkpoint in Nilgiris District; Puliyarai Checkpoint, Tirunelveli District; Lower Camp Checkpoint, Periyar Project in Madurai District; and Kaliakavilai Checkpoint in Kanniyakumari District.

78. http://www.tn.gov.in/sta/contacts–c.htm (Accessed December 2006).
79. www.tnsalestax.com/OFFICERS%20NAME%20IN%20CHECKPOST.pdf, http://www.tnvat.gov.in/English/CT%20at%20A%20Glance_2006–07.pdf (Accessed December 2006).
80. Velayutham Saravanan, Decline of Tribal Economy in Tamil Nadu: 1947–2000 in Govinda Chandra Rath (ed), *Tribal Development in India: The Contemporary Debate*, New Delhi: Sage Publications, 2006, pp. 218–236; Velayutham Saravanan, 'Economic Reforms and Tribals in Tamil Nadu: 1991–2003' *Social Action,* 56(1), 2006, pp. 25–39; Velayutham Saravanan, 'Economic Transformation of Tribal Women in Tamil Nadu: 1947–1997', *Indian Journal of Labour Economics,* 42(4), 1999, pp. 699–706; Velayutham Saravanan, 'Economic Transformation of Tribals in Tamil Nadu: Causes of Migration and its Employment Pattern, 1792–1991', *Indian Journal of Labour Economics*, 41(4), 1998, pp. 695–705; Velayutham Saravanan, 'Impact of Welfare Programmes on the Tribals in Tamil Nadu: 1947–1991,' *Review of Development and Change*, 3(1), 1998, pp. 90–108; Velayutham Saravanan, 'Economic Transformation of Tribals in Tamil Nadu: Subsistence to Commercialisation 1976–1995,' *Review of Development and Change*, 2(1), 1997, pp. 157–168.

CHAPTER 6

Impact of Forest Rights Act, 2006

In an electoral democracy that is held hostage by populist politics, tribal development in post-colonial India has by and large been viewed from the angle of generating a vote bank and not from a developmental point of view. It took many years to acknowledge the truth that restrictions on the tribal use of forest resources, imposed as a tool for conservation, were a historical injustice. When, during the late 1970s, tribal development emerged on the agenda of mainstream politics, some efforts were made to provide infrastructural facilities, but the situation did not witness significant change. Recent legislation, in an attempt to correct historical wrongs, ensures tribals can access a hitherto forbidden area, the very forests in which they live. The interplay of power and populist politics has had an unmistakable role in affecting both historical restrictions as well as the currently claimed revision. Since the core issue of tribal rights has remained largely unaddressed, it seems that even after fifty years of Indian planning, the deprived groups, particularly tribals, continue to remain underdeveloped, with their living conditions deteriorating further.

This chapter raises the pertinent question of whether or not the new legislation provides an impetus to the economic progress of India's tribals and at the same time enhances environmental sustainability. The chapter tries to understand the different undercurrents beneath the post-colonial

An earlier version of the chapter was published in *South Asia Research*, 29 (3), 2009.

© The Author(s) 2018

V. Saravanan, *Environmental History and Tribals in Modern India*,
https://doi.org/10.1007/978-981-10-8052-4_6

approach towards tribals along with the underlying connotations as well as the implications, both regarding the environment and tribal development, in the context of India's latest statutory development in this field, the Scheduled Tribes and Other Traditional Forest Dwellers (Recognition of Forest Rights) Act 2006.

On the one hand, a skewed understanding of environmental protection has thus far curtailed the growth of the forest-oriented tribal economy. Restrictions imposed through the different Forest Acts from the late nineteenth century onwards continued to operate during the post-independence period. On the other hand, tribal development policies gave an impetus to the creation of important infrastructural facilities. Bringing roads, communications, schools, hospitals and credit facilities to remote areas, they claimed to be attempting to integrate the economy of the tribal communities with mainstream society. The Tribal Sub-Plan (TSP) of the last two decades of the twentieth century in particular aimed to accomplish this agenda. However, these measures, intended to protect the environment and to accentuate the pace of development initiatives, have largely failed. Environmental protection was not ensured, and neither were there any dramatic improvements in the living conditions of tribals in many parts of the country.

In the context of the political economy, the new Recognition of Forest Rights Act 2006 has to be subjected to scrutiny. Taking a historical perspective, the political undercurrents of economic policymaking towards the tribals in the context of tribal development planning in the 1980s are examined, before analysing the likely impact on tribal development and the environment of the Recognition of Forest Rights Act 2006.

Tribal Development and Environment

Though the focus of this chapter is on the contemporary period, the core issues concerning conservation and tribal development cannot be seen in isolation from their colonial legacy. Since the late nineteenth century, blaming tribals for the denudation and destruction of the green cover has been the standard rationale for imposing severe curbs on their use of forests. The colonial government imposed several restrictions under various Forest Acts to protect and preserve the environment. Under these Acts, during the late nineteenth century, to a very large extent the forests used by indigenous tribal communities without any interference from external agencies such as Forest Departments were notified as reserve forests. While imposing these restrictions, the colonial government did not initiate development programmes in the hill regions.[1] The same period also witnessed official encouragement by the Forest Departments to commercialise forest resources in different parts of the country.[2] Early post-colonial governments, then, did not differ much from their predecessors, either in terms of imposing controls on tribal access to forest resources or in carrying out development measures.[3] Indeed, tribal development continued to be largely ignored in the absence of concrete programmes with adequate financial allocations.

However, this trend changed particularly after the 1980s, when the TSP concept was introduced during the Fifth Five-Year Plan. From that point, certain infrastructural facilities came to be established in tribal areas. Unfortunately, however, these initiatives have failed to bring any substantial improvements and the living condition of tribals has deteriorated further, with a pathetic decline in their status, often from cultivators to labourers.[4] Interestingly, the post-colonial governments also continued to encourage the commercialisation of forests.[5] Accepting the failures of forest policy, the Ministry of Tribal Affairs began to critique restrictions on the tribals as historical injustice and sought to ensure access to forest resources, besides conferring ownership rights for the lands under their possession. However, curbs to access forest resources were imposed wherever wildlife was affected.

Initial resistance to the new law in 2005, first in the shape of a bill laid before Parliament, came from the Ministry of Environment and Forests, indicating that it would hinder efforts to preserve forest resources. It was pointed out that the new legislation would require the de-notification of vast tracts of forests and the elimination of legal protection for forest cover, which would lead to colossal ecological damage. Indeed, the Ministry had earlier initiated steps to regularise claims under dispute, and already in 1990 had issued six circulars in this regard.[6] According to these circulars, pre-1980 encroachments on forest lands were considered eligible for regularisation, provided the state governments had evolved certain eligibility criteria in accordance with local needs and conditions. But the state governments did not respond to these circulars.[7] Meanwhile, the judiciary was approached to stall the process. To vacate the Supreme Court's stay on the February 2004 Circular to extend the date of regularisation to 1993, the Ministry of Environment and Forests stated that rural people, especially tribals who had been living in the forests since time immemorial, were deprived of their traditional rights and livelihood and, consequently, the tribals had become encroachers in the eyes of the law. It further stated that it should be understood clearly that lands occupied by tribals in forest areas did not have any forest vegetation, and that regularisation would significantly lead to better forest conservation.[8] Did this signal a change of mind and policy?

Forest Policies: An Appraisal

The Scheduled Tribes and Other Traditional Forest Dwellers (Recognition of Forest Rights) Act 2006 marks a clear paradigm shift in the approach of the Ministry of Environment and Forests towards the Scheduled Tribes, who were hitherto held responsible for the destruction of forests. So far the plight of the tribals, whose economy was associated with forest resources, had been largely ignored and disregarded with utter contempt. It is a well-established fact that since the early nineteenth century forest resources had been increasingly extracted for commercial purposes, while restrictions were clamped on the indigenous tribals through a number of Forest Acts and a plethora of rules and regulations.[9] Several strategies, including force, were employed to control the forest areas to facilitate commercial exploitation.[10] For example, in Madras Presidency, the Forest Department first brought vast tracts of forest land under forest reserve by the close of the nineteenth century. Later the Revenue Department initiated survey settlement during the early twentieth century.[11] This was an exercise whereby the Forest Department took over large chunks of forests, an inalienable part of the tribals' livelihood, and then handed the local tribals the remaining areas with ownership rights. While part of the source of their livelihood was thus made inaccessible, the rest was permitted to be utilised with an eye on revenue generation.[12]

After independence, forest resources continued to be extracted for commercial purposes and increasingly to meet industrial needs. In addition, various developmental projects such as dams and activities related to tourism consumed a very large extent of the green cover.[13] For example, in 2001 alone about 414 paper industries depended on forests for the supply of raw materials. There were about 41,229 sawmills in that year with a potential capacity of 29,564 thousand cubic metres per year. In addition to this, about 407,983.02 hectares of forest land were diverted for non-forest use between 1981 and 1998.[14] According to these Forest Department statistics, in 1998 about 714,855.30 hectares of forest land were under encroachment.[15] According to the Ministry of Environment and Forests, the total area of forest land under encroachment by both tribals and others was about 13 lakh hectares.

Several states enacted laws to curtail tribal land alienation, but invariably they remained ineffective.[16] In what is a clear case of locking the stable after the horses had fled, several states enacted relevant laws only when the issue of land alienation had reached its peak. Consequently, one could find a large number of non-tribal settlements in the hill/tribal areas.[17] The logical corollary was a massive displacement of tribals, who were pushed into positions as labourers on their own lands, indicating a deterioration of conditions in the hills.[18] The post-colonial forest policy, specifically the National Forest Policy 1952, had its thrust and focus on the uninterrupted supply of forest resources primarily to satisfy the needs of industry and other commercial purposes. Following in the footsteps of the past, the post-colonial government also continued to place blame on the tribals for the deforestation and depletion of forest resources and tightened its restrictive grip, either through existing Acts or the enactment of new legislation until 1988. In the history of forestry in the Indian subcontinent, tribal people's dependence on forests was first recognised in the National Forest Policy of 1988.[19]

State intervention and attendant mechanisms for development also proved to be futile.[20] The depletion of flora and fauna along with a decline in the standard of life of hill inhabitants were all very much evident. Consequently, the realisation gradually dawned that without the active cooperation of the tribals, the conservation of forests and wildlife would be a difficult task. Recognising the symbiotic relationship between tribals and forests, the National Forest Policy of 1988 made provisions to safeguard the customary rights and interests of the tribals and forest-dwellers over forest resources.

The Ministry of Environment and Forests had laid strong foundations for the 2006 Act almost two decades earlier. Based on the National Forest Policy of 1988, the Ministry had issued several circulars in 1990 to regularise the unauthorised lands held by the tribals in forest areas.[21] However, the state governments failed to act on these guidelines.[22] In addition to this, in the same year the Government of India outlined a framework for creating a massive people's movement through the involvement of village committees to protect, regenerate and develop degraded forest lands. Conscious of this, the Forest Department also initiated confidence-building measures through the Joint Forest Management programme.[23] The last two decades also witnessed the issuance of several circulars with a focus on conservation, but unfortunately these measures were largely confined to degraded forests. As such, the 2006 Act was to some extent an admission of the growing problems and plight of the tribals.

Political Economy of Tribal Development

While placing restrictions on the use of forest resources during the colonial period was not accompanied by any corresponding development initiatives for the tribals' socio-economic progress, the same approach continued in the early post-colonial era despite several constitutional safeguards. In the absence of concrete programmes, these safeguards obviously lacked credibility. Some half-hearted measures were taken up in the hill and jungle areas as part of the rural development programmes prior to the TSP. These failed to succeed, mainly because of the isolation of the tribals and a lack of communication facilities in their areas. Developmental plans and programmes targeting tribals were initiated only during the Fifth Five-Year Plan with the launch of the TSP, which was a milestone.

The TSP was implemented with the twin objectives of assisting Scheduled Tribe families to cross the poverty line and to provide basic amenities and facilities in their areas. The plan was thus conceived to help tribal families to improve their socio-economic and educational status. Economic development programmes in the fields of horticulture, animal husbandry, soil conservation, minor irrigation, sericulture, small industries and bee-keeping, besides locale-specific schemes for the provision of drinking water, road connectivity, electrification, education and health facilities, improving forestry, communication, cooperation and social services were all taken up. Whether these measures created the desired impact in the economic sphere requires closer scrutiny.

According to the 2001 Census, the total population of Scheduled Tribes stood at 84,326,240, roughly constituting 8.2% of India's total population.[24] Of this, 91.7% lived in rural areas, while the remaining minority of 8.3% resided in urban areas. That the development initiatives failed to create a positive impact was very much visible through a decline in the proportion of tribals' participation rates in economic activities. According to 2001 Census figures, the work participation rate among the Scheduled Tribes was as low as 49.1%. About 69% of these were classified as main workers, while the remaining 31% were treated as marginal workers. Decline in the status of such workers was also evident from the fact that no significant structural transformation had occurred among the tribals. About 82% of the tribal workforce was engaged in primary sector activities (cultivation and agricultural labour). If agriculture and allied activities were included, this proportion might increase. At the same time, the proportion of cultivators to the total tribal workforce declined to 45%, while the category of agricultural labourers registered an alarming increase of 37% in 2001.[25] More or less the same trend prevailed in different parts of the country.[26] This provided a clear indication that the tribal economy was on a downward slide, with a visible downfall owing to large-scale encroachment by the Forest Department and alienation of land by non-tribals.[27] For instance, in Tamil Nadu the Forest Department has added more areas to the reserve forest since independence. Consequently, the area under forest cover has increased from 1,866,000 hectares in 1960/1961 to 2,134,000 hectares in 1999/2000, or from 14.13% to 16.55% of the total geographical area of the state.[28] A survey undertaken among 2631 tribal households in 1986 to ascertain the intensity of land alienation revealed that 37% of the tribal households had lost their lands to outsiders.[29] Land was either sold or disposed of to meet rising domestic consumption, debt repayment, social and religious ceremonies, medical expenses and to some extent alcoholism.[30] Further, illiteracy remained very high. According to the 2001 Census, about 53% of the total tribal population in India was classified as illiterate; the proportion was very high among women, at 65%.

Given the declining trends in the tribal economy, the impact of development programmes providing safe drinking water, electricity and toilet facilities needs to be examined. According to the 1991 Census, only a meagre 3.23% of tribal households had access to safe drinking water, electricity and toilet facilities. A tiny 5.02% of households had both electricity and toilet facilities. Needless to say, almost all the tribal households, especially in the rural areas, were living without the basic minimum facilities, as shown in Table 6.1.

However, it could not be ignored that some facilities were made available to a sizeable section of tribal households. For instance, 43.21% of households had access to safe drinking water (a tap, handpump or borewell), electricity (22.8%) and toilet facilities (7.22%). Interestingly, however, about 45.3% of the tribal households did not have any of these basic facilities. According to the 2001 Census, about 63.5% of households and 47.8% of villages in the tribal belt had no electricity. Further, only about 17% of Scheduled Tribe households had lavatories within their premises. What emerges from the above data is that basic amenities remain woefully inadequate in the hills. There are about 2690 forest villages in thirteen states which are still managed by the Forest Departments of the respective state. These villages are outside the Revenue Administration of the districts and have therefore missed the fruits of development.[31] Other tribal

Table 6.1 Sources of facilities available for tribal households, 1991

Facilities	*Total*	*%*
Safe drinking water[a]	5,335,483	43.21
Electricity	2,814,885	22.80
Toilet facilities	892,116	7.22
Safe drinking water, electricity and toilet available	398,830	3.23
Electricity and toilet available	620,324	5.02
Safe drinking water, electricity and toilet not available	5,593,598	45.30
Electricity available and toilet not available	2,194,561	17.77
Electricity not available and toilet available	271,792	2.20
Total	12,348,000	

Source: Census of India, 1991, *Tables on Houses and Household Amenities*

[a]Tap/handpump/borewell

villages, too, are lacking basic infrastructural facilities. Shah notes: 'In the absence of a secure title to land, adivasis [tribals] are deprived of their rights as farmers—unable to access credit, electricity or agricultural inputs and deprived of benefits of various anti-poverty programmes. Since their names do not exist in the land records, they cannot become members of tribal cooperative societies.'[32]

The severity of penalties for 'offences' and restrictions on the use of forest resources, including the collection of firewood, left the hill inhabitants with very few alternative sources even for cooking. This is confirmed by examining the sources of fuel for cooking. This does not require much explanation, since Table 6.2 clearly indicates that more than 97% of the tribal households were dependent upon firewood in 1991.

The trend would not be different even if the 2001 Census data was considered. As the availability of day-to-day livelihood needs has come under increasing stress, many tribals are facing cases filed by the Forest Department, besides being subjected to harassment. This amounts to the violation of their right to livelihood, an integral part of human rights. Most of the tribal households are trespassing into the forests in one way or the other, at least once a week, for the collection of firewood. A large number of registered cases fall under the category of unauthorised felling, grazing and other minor offences.[33] Given this situation, the 2006 Act facilitates the tribal use of forest resources without facing any kind of threat and harassment from the Forest Department, and this is helpful.

Table 6.2 Type of fuel used for cooking in India, 1991

Type of fuel used for cooking	*Total*	*%*
Cow dung cake	632,835	5.13
Electricity	13,756	0.11
Coal/coke/lignite	195,360	1.58
Charcoal	54,034	0.44
Cooking gas	132,479	1.07
Wood	10,805,582	87.51
Bio-gas	20,101	0.16
Kerosene	238,101	1.93
Others	253,707	2.05
Total	12,348,000	

Source: Census of India, 1991, *Tables on Houses and Household Amenities*

However, one can safely draw a conclusion here that the development initiatives have not so far effected a positive transformation in the livelihood patterns of tribal households. The government itself admits: 'The gap in the infrastructure in the tribal areas vis-à-vis the rest of the areas is further widening at much faster rate. For example, while on the one hand the quality of roads, health services, telecommunications, distribution of power, etc. are improving in the country through participation of private sector, the condition is deteriorating in the tribal areas due to even poor maintenance of already created assets.'[34]

Since 2005, however, there has been a discernible shift in the official discourse on tribal development. Has the development paradigm been turned upside down? Giving up the rhetoric of blaming tribals for deforestation, the government has started to argue the merits of allowing them to use forest resources. Is this just a ploy to evade the responsibility for the unsuccessful development policies that have been pursued for three decades? Even while denying access to forests, developing infrastructure facilities in tribal areas was on the agenda of politics in the 1980s. Now it appears that sending the tribals back to the forests is the most recent slogan. It is appropriate to quote Singh: 'It should shame us that this is all that India, the future economic superpower, can offer its most desperate citizens in the twenty-first century. It should shame those who rule us that they have been able to do so little by way of development in nearly sixty years of independence that they offer our supposedly beloved 'tribals' stone age living standards. But they are not ashamed; they are proud.'[35]

Features of the Recognition of Forest Rights Act 2006

Indeed, after almost six decades of independence, for the very first time the government has realised the historical wrong committed towards the tribals and recognised their plight as well as their increasing marginalisation from forests because of the different Forest Acts, Rules and policies. It should be pointed out that the parties in the ruling coalition at the Centre, the Congress-led United Progressive Alliance (UPA), lost their vote share among tribals in different parts of the country. It appears that to regain their support and to retrieve lost ground, the government put forward the 2006 Act. As pointed out by Rangarajan, 'The call for a fresh look at the issue of adivasi land rights did not spring from a vacuum but was rooted in the desire of the Congress and allied parties to recover lost ground among an increasingly assertive section of the citizenry.'[36]

Hence, this agenda was included in the UPA's National Common Minimum Programme. Based on this programme, the Government of India decided to introduce this legislation, and included it in the Presidential address.[37] In his customary address to Parliament, which enunciated the government's priorities, the President stated on 25 February 2005: 'Adivasis in this country have been leading a life of insecurity in many areas since their property rights have remained unsettled. It is necessary to recognise their need for assured property and land rights in areas they have been residing in for several generations. It is an irony that tribals who have been living in "forest villages" and have been practising agriculture on these lands for several generations, have not been given due recognition of their rights. Their problems are engaging the attention of the Government and we will try to settle the issue of land rights of tribals. The outcome will be beneficial both to tribals and to the goal of forest conservation.'[38]

Accordingly, the government introduced the Scheduled Tribes (Recognition of Forest Rights) Bill 2005. The objective of the Bill was stated as '[t]o undo the historical injustice by recognizing and vesting the forest rights and occupation of forest land to forest-dwelling Scheduled Tribes who have been residing there for generations and who are integral to the very survival and sustainability of the forest eco-system, including wildlife, but whose rights could not be recorded'.[39] The Draft Bill proposed to give tribals several things: 2.5 hectares of land per family, ownership rights to access minor forest produce, the right to graze and access to traditional seasonal resources.

Protest Over the Bill

The participants in debates on the Scheduled Tribes (Recognition of Forest Rights) Bill of 2005, prepared by the Ministry of Tribal Affairs, could be broadly grouped into three categories: tribals and other forest-dwelling activist groups supported by the Ministry of Tribal Affairs who argued in favour of the bill; environmental/conservation activists dominated by the wildlife (particularly tiger) protection groups supported by the Ministry of Environment and Forests who opposed the bill; other forest-dwelling groups assisted by civil society organisations and certain political parties who supported efforts to include other forest-dwelling communities.[40]

Against the proposed bill, several criticisms cropped up that could be broadly grouped into three categories. The conservationists, environmentalists and wildlife activists, especially the pro-tiger enthusiasts, were in the forefront of opposition to the bill. Their contention was that the proposed legislation would cause irreparable damage to wildlife. Expressing apprehension that it would result in the transfer of 50 million hectares of the country's 67 million hectares of forestland to tribals and forest-dwellers, they pointedly called the Act a 'sell-out to vote-bank politics'.[41] It must be noted, though, that the argument put forward by the tiger conservationists was due to a misreading of the provisions in the bill. Hypocrisy is thus rightly identified in the attack on the bill by conservationists and some well-known journalists.[42]

Apart from these groups, the Ministry of Environment and Forests feared that this new law would affect forest cover, arguing that the approach adopted in the proposed bill requiring de-notification of vast tracts of forest lands and the elimination of legal protection for forest cover would lead to irreparable ecological damage of immense proportions. Further, it was said that, as reported by the *Indian Express* on 16 April 2005: 'Failure on the developmental front should not be compensated by any mechanism which leads to the irreparable damage of our natural resource base. The draft bill, inter alia, proposes to compensate the failure on the developmental/welfare fronts by distributing natural resource base of the country.'

The principal objections of the Ministry of Environment and Forests were that giving power of settlement claims to village councils (*gram sabhas*) would result in local vested interests gaining control; the distribution of 2.5 hectares of forest land to each family was against the goal of the

National Forest Policy of 1988 to bring one-third of the country under tree cover; the denotification of vast tracts of forest lands and elimination of legal protection for forest cover would have disastrous consequences on the ecological balance; and over 60% of India's forests would be handed over to 8.2% of its population.[43] Some of the tiger protectionists proffered dramatic solutions for tribal development.[44]

Countering the environmentalist groups, the tribals and other forest-dwelling groups submitted that the bill would not distribute land, but would merely settle claims of people already living on those lands. Further, existing legal protection for forest cover would not be eliminated, since the Bill insisted upon forest rights' holders not only to refrain from 'any activity … that adversely affects the forest and the biodiversity in the local area', but also required the entire community to 'stop any activity which adversely affects wildlife, forest and biodiversity, whoever is responsible'. In fact, this was the first law that actually required the involvement of local communities to protect the forest.

Dismissing the claim of the environmental groups that the existing statutes contained enough guarantees for people's rights, it was argued that except for the Panchayats (Extension to Scheduled Areas) Act of 1996, and a set of Ministry of Environment and Forests circulars in 1990, which as we saw proved to be unsuccessful, other Acts would not help to regulate the issue of land rights. The blame for the degradation of forests was solely attributed to the timber mafia, land grabbers, industry (particularly mining) and the diversion of a large proportion of the green cover for non-forest use. According to this lobby (tribal group), 3.33 million hectares of natural forests were cleared and replaced by industrial plantations between 1951 and 1979. From 1980 onwards, the forest authorities had diverted 9.81 lakh hectares of land for non-forest use, particularly for 11,282 development projects, but completed only 7.3% of the required compensatory planting.[45] Owing to harassment by Forest Department staff, the tribals suffered so much that the new legislation had become necessary to protect them.[46] The third group of lobbyists, other forest-dwelling groups, argued that the demand for the inclusion of forest-dependent Scheduled Castes and Other Backward Communities gave room for further controversy.[47]

The bill was referred to the Joint Parliamentary Committee on 13 December 2005. This suggested a change in nomenclature of the draft bill to Scheduled Tribes and Other Traditional Forest Dwellers (Recognition of Forest Rights) Bill 2006. Further suggestions were to include the communities who had been traditionally living in or adjacent to forests for at least three generations; that those who had occupied forest land before 13 December 2005 should be considered for recognition and vested with forest rights; and that each family should be entitled to 4 hectares of forest land rather than 2.5 hectares as suggested in the bill.

On 23 May 2006, the Committee presented the revised Scheduled Tribes and Other Traditional Forest Dwellers (Recognition of Forest Rights) Bill, 2006. Its suggestions were accepted by the government and the bill was finally passed in the winter session of Parliament, with presidential assent granted on 29 December 2006. The Act was published in the Gazette of India, Extraordinary on 2 January 2007.

The Provisions of the 2006 Act

This chapter attempts to situate and contextualise the Scheduled Tribes and Other Traditional Forest Dwellers (Recognition of Forest Rights) Act 2006, arguing that this new law will have far-reaching implications on the environment as well as tribal development. The Act explicitly allowed tribals to use forest resources for their livelihood, hitherto restricted by the earlier Forest Acts since the late nineteenth century. For instance, the Indian Forest Act 1927 prohibited any fresh clearing for cultivation and other purposes in the notified areas, cattle trespass and grazing, felling of trees and damaging of any tree besides hunting activities in the reserve forests. Under this Act, any violation of the rules would be punishable either by imprisonment or penalty or both. In addition, damage to property had to be compensated. In contrast, the 2006 Act endowed the tribals with the right to live in the forest, the right to cultivate for their livelihood, the right to collect minor forest produce, the right to graze cattle, the right to convert leases or grants (*pattas*) to titles, the right to convert forest villages into revenue villages, the right to settlement in their old habitations and unsurveyed villages, the right to access and community rights over intellectual property and traditional knowledge related to forest biodiversity and cultural diversity, the right to manage the community forest resources and the right to enjoy any customary/traditional practice, excluding hunting.

The legislation accorded due statutory recognition to what existed prior to the erstwhile Forest Acts and external interventions into the self-sustainable forest-based livelihood systems and traditional customary practices, within the purview of the existing forest and wildlife acts. It ensured ownership rights to those tribals who did not have proper documents despite having lived in the hills for generations. In most states, the Forest Departments notified vast tracts as forest reserve before granting ownership rights to the tribals for the land under their possession. For instance, in Madras Presidency reserve forests were notified in hill areas in the last decade of the nineteenth century, while the survey and settlement were held during the first decade of the twentieth century.[48] Consequently, the tribals, who had no occupancy rights, were treated as encroachers on their own land and lived under the constant threat of being evicted or penalised. According to official Forestry Statistics, about 6.17 lakh hectares of forest land were classified as encroachments.[49] Setting right the anomaly, the Act would ensure removal of bottlenecks in the implementation of the welfare measures, which had been previously denied to so-called 'unauthorised settlements'.

Significantly, the new legislation was expected to drastically bring down the number of forest-related offences as well as harassment faced by the tribals and other forest-dependent communities. Illicit felling and grazing still account for the bulk of offences. According to Forestry Statistics, about 159,094 cases were registered and Rs.8303.47 lakh was recovered during 1996/1997 in the illicit felling category alone. Illicit grazing accounted for 37,190 cases during the same period.[50] The dismantling of decades-old barriers and legal hurdles would eliminate harassment at the hands of authorities and should greatly relieve tribals from the fear of officialdom.

Conflicts between Environment and Tribal Development

At the close of the nineteenth century, a large proportion of the present forest cover was brought under the reserve forests system, which negated the customary rights of tribals. Naturally, a series of conflicts ensued between the hill communities and the Forest Department while the lands under the former's occupation were demarcated and over access to natural resources. Several representations were made over time to exclude tribal land from the proposed forest reserves.[51] Ultimately, the concept of reserve forest emerged to become a part of forest management and conservation.

The tribal population density in the hill areas even a century ago was very low. As such, enormous amounts of land in different regions remained unoccupied. This facilitated the bringing of more and more forest terrain under direct state control. Owing to a lack of communication and transport facilities, Forest Department staff mostly remained at villages in the foothills, while the tribals continued to use forest resources for their livelihood. Non-tribal settlements were initially confined to a few hills, where plantations were established.[52] The advent of this plantation sector added another dimension to existing conflicts, and sharpened them, especially with the increasing pace of land alienation. Development of infrastructure such as communication, transport and road connectivity in the hills during the post-independence period assisted the Forest Department in mounting increased vigilance.

In this context, a pertinent question is how the new Act of 2006 would address the issue of tribal development and environment, always seen as conflicting rather than being complementary. The constitutional position on this is actually very clear. Article 48A of the Constitution of India of 1950, added by the 42nd Amendment in 1976, provides that '[t]he State shall endeavour to protect and improve the environment and to safeguard the forests and wildlife of the country'. Article 46 of the Constitution further seeks to promote the interests of weaker sections of the population and provides that: 'The State shall promote with special care the educational and economic interests of the weaker sections of the people and, in particular, of the Scheduled Castes and the Scheduled Tribes, and shall protect them from social injustice and all forms of exploitation.'

In pursuance of such clear constitutional mandates, included among the Directive Principles of State Policy in Part IV of the Indian Constitution of 1950, several statutes were enacted to protect and preserve environment and wildlife. Despite this, there was an alarming depletion of natural resources and wildlife species in India. With regard to tribal development, the state implemented several programmes, though they have failed to contain exploitation.

Unfortunately, the 2006 Act appears to be contrary to the Constitution in two ways. First, attempting to recognise centuries-old, but now defunct, forest-oriented customary rights could turn out to be a great threat to the environment, ecology and wildlife. Secondly, instead of assisting and facilitating the tribals' economic progress and protecting them from exploitation, encouraging them to live in the forests could be interpreted as amounting to an evasion of responsibility.

In environmental terms, the existing green cover is far below the requisite proportion to the total geographical area of the country. At present, forests account for only about 20% of the total area of India, and the actual green cover could be even less. According to the norm set by international agencies, about one-third of the total geographical area should be covered by forests; likewise, the National Forest Policy of 1988 envisaged an increase in forest cover to 33%. As far as the hills and mountain regions are concerned, the National Forest Policy has suggested maintaining two-thirds of their area under forest cover. The Tenth Five-Year Plan document also indicated that this 33% green cover could be achieved by the end of the Eleventh Five-Year Plan period.

The main apprehension of environmentalists, as noted, is that the 2006 legislation, if implemented, would pave the way for handing over roughly 60% of the present forest area to tribals. It is true that not necessarily all tribal households would get land from the forest, but the fact remains that all tribals and other forest-dwellers are entitled to collect forest produce. Forest resources in their entirety have thus been virtually given back to the tribals and other forest-dwellers. Since denudation was rampant when there were restrictions on collecting forest resources, the consequences could be dangerous once unfettered collection becomes a matter of right. It is apparent that de-reserving forest reserves and allowing the collection of forest produce, in the context of shrinking forest cover, could pose a serious threat to the ecosystem, already under severe stress from various actors. This raises the important question of how responsible tribals will feel for protecting the green cover of their living environment.

Admittedly, a shift from primary sector activities to those of the secondary and tertiary sectors is a better indicator of economic transformation of any marginalised community or developing nation. Policies with this objective are invariably being pursued by different countries to attain faster growth and development. As far as India's tribals are concerned, a shift from forest-based subsistence economy to other economic activities becomes a prerequisite for their socio-economic transformation. Accordingly, the government has adopted strategies to integrate tribals within mainstream society by providing necessary infrastructural facilities since independence, particularly from the Fifth Five-Year Plan onwards. Unfortunately, the tribals were not integrated within the national mainstream, and no tangible progress has been made on the economic front in hill areas. Instead, an influx of non-tribals into these areas and the consequent land alienation has continued unabated, pushing the indigenous tribes further into the inaccessible remote forests or turning them into dependants on non-tribals in and on their own land. Given this particular situation, it is doubtful whether granting forest rights can produce a positive transformation among the tribals; rather, it might prove to be an attempt to push them still further backward by empowering the non-tribal encroachers. Certainly, this Act would be helpful to people who occupy land without land rights. But what about the landless labourers who account for about 39% of the tribal population? Will providing forest rights help them? It appears that this is yet another facet of popular politics which does not hold any hope for tribals' economic progress.

It is a known fact that the tribals only had occupancy rights before the survey and settlement era; at the time when reserve forests were notified, the Forest Department questioned the validity of occupancy rights. Even after the survey and settlement, when ownership rights were granted, the non-tribals alienated those lands on a large scale. Non-tribal settlements have thus increased in the tribal belts, particularly during the post-independence period. As a consequence, as noted, many tribals have become landless labourers and to some extent 'encroachers' in the forest areas. Though some states have enacted laws to prevent tribal land alienation, the desired results are not yet in sight.

At present, there are 194 Integrated Tribal Development Projects, 254 Modified Area Development Approach programmes and eighty-two clusters spread over twenty-one states and two union territories. In every project area, a sizeable proportion of non-tribal settlement remains visible. In the context of continuing land alienation, what guarantee is there that ownership rights for the 'unauthorised' occupied lands of the tribals would prevent alienation in future? Moreover, this Act now confers legitimacy on the settlements and alienated or occupied lands of non-tribals. As such, it will actually sharpen the problem of land alienation. Despite the restrictions in force, non-tribals have managed to circumvent the rules and have alienated tribal land. There is much evidence that powerful non-tribals continue to engage in this, adopting different strategies. Hence, treating non-tribals on a par with tribals militates against reason. Instead, if the state had taken decisive steps to restore the lands alienated from the tribals themselves, this would have proved to be more viable, probably in terms of conservation as well.

Another significant feature of the 2006 Act is the emphasis on restoration of traditional tribal institutions, which were among the first casualties of colonial intervention, particularly during the early nineteenth century. For example, in the different hill regions of Madras Presidency, the colonial government crippled the functioning of tribal institutions. The modern village administrative system with its elected representatives replaced the earlier community institutions. States such as Tamil Nadu took steps to curb the traditional village *panchayats* in the plains, which seemed to be an anachronism in the rule of law. Moreover, in different parts of the country the traditional tribal institutions either remain dead or have become ineffective. As rightly pointed out by Burman, 'gram sabhas in most parts of the country are ineffective and their existence totally depends on the *panchayat* bodies and government functionaries at the grassroot'.[53] The existing democratic institutions at local level are providing a much wider space than the earlier traditional system, not only in terms of electing members and replacing hereditary lineage, but also by being instrumental in introducing various development programmes. In addition to this, they provide checks and balances through accountability, for there are elections at least once every five years. In these circumstances, restoring institutions which disintegrated almost one and a half centuries ago could at best be plain nostalgia and at worst undesirable. Further, as Bhatia notes, 'the actual powers of the gram sabha, and the relation of the gram sabha to other authorities, are far from clear'.[54]

Infrastructure development in the hills received priority after the Fifth Five-Year Plan, with the state drawing up several development strategies to expedite the pace of the tribal communities' socio-economic transformation. However, the failure of the state to arrest non-tribal settlements left the alienation of tribal lands to continue unhindered. Ultimately, the tribals were either pushed into reserve forests or became dependent on others. Hence, unless the state restored these alienated lands from the non-tribals and ensured better delivery systems, no improvement in the tribals' living conditions could be expected.[55] Past experience shows that development facilities have only paved the way for further displacement of the tribals.

Under the draft bill, the cut-off date for the granting of rights was 1980. The Joint Parliamentary Committee recommended a final cut-off date of 13 December 2005, which was accepted by the government. This means that those who had occupied lands prior to 2005 would become eligible. Here, a comparison with the colonial past becomes inevitable. Subsequent to the introduction of reserved forests during the late nineteenth and early twentieth centuries, the survey settlement conferred ownership rights on tribals for their lands under cultivation.

If there is any dispute between the Forest Department and tribals at the time of notification regarding the occupied lands, there is a presumption that ownership rights will be provided. Taking into account the 'unauthorised' tribal occupants, however, the restoration of alienated tribal lands has been conveniently ignored.[56] The significance of this cannot be dismissed. Hence, along with tribal ownership rights to the land under their hold at the time of forest notification, restoration of the alienated lands after settlement should be done simultaneously to prevent further land alienation and protect the environment and ecology.

Suitable amendments to the Forest Acts of both central and state governments remain a pre-requisite for any corrective mechanism to yield the desired results. While the bureaucracy in the police and revenue departments continue to play a role in maintaining a restrictive and coercive regime, the Forest Department holds the key in many respects.[57] According to Section 12 of the Forest Act 1927, '[i]n the case of a claim to rights of pasture or to forest-produce, the Forest Settlement-Officer shall pass an order admitting or rejecting the same in whole or in part'. The Wild Life (Protection) Amendment Act 2002, Section 34A(a), authorises eviction of 'any person from a sanctuary or National Park, who unauthorisedly occupies Government land in contravention of the provisions of this Act'. It has been further pointed out that 'section 15 can effectively nullify the positive provisions of the Act through the very colonial legislation that the Act is supposed to rectify'.[58] Hence, making appropriate changes in the various Forest and Wildlife Acts to make them more compatible with the 2006 Act assumes very great importance.

Finally, the Act also attempts to ensure the tribals' intellectual property rights. Knowledge was never documented among the tribal communities, who passed it on orally to the next generation. The state's restriction of tribal access to forests, which was central to their social and cultural life, resulted in a critical disjunction. The generations after colonial intervention are obviously handicapped in terms of acquisition as well as the transfer of traditional knowledge. Unlike their forefathers, they did not have unfettered access to the forests, and their knowledge levels gradually declined. The fact is that the intellectual property of the tribals has been lost over a period of one and a half centuries; attempts to retrieve it would prove futile. Thus this Act would neither benefit the tribal communities nor enhance conservation, but would promote only the interests of the non-tribals.

Conclusions

Development has been broadly accepted to mean a process of improving people's status and living conditions, while conservation denotes the protection of natural resources. The present new legislation might facilitate tribals' use of forest resources for their subsistence and guarantees them ownership rights. At the same time, it has to be borne in mind that the 2006 Act is silent on the restoration of lands from non-tribals that were occupied prior to the cut-off date. The entry of non-tribals into the hills preceded the establishment of the Forest Department in the early nineteenth century. Even after the introduction of the reserve forests during the late nineteenth century, a very large extent of tribal lands was alienated to non-tribals. In addition, a sizeable extent of tribal land was alienated by educational institutions, hotels, resorts and other tourism development ventures. Consequently, about 37% of the tribal population in India had become agricultural labourers by 2001. No one can possibly argue that tribal households were excluded from forest lands prior to the intervention of the Forest Department or the advent of non-tribal settlements. Every household occupied the land for their own survival. While non-tribals cornered most of the fertile and cultivated lands, the tribals were forced to go deeper into the forests or become dependants. Therefore, restoring alienated land is a more viable and sustainable option than just giving tribal people ownership rights over forest lands.

From the development perspective, as indicated earlier, a shift from primary sector to secondary and tertiary sector activities is a better indicator of economic progress. In tune with this, governmental policies have provided the requisite thrust for improving infrastructural facilities in tribal areas, particularly during the last two and a half decades of twentieth century. However, it has made very little impact on their socio-economic development, primarily because of ineffective delivery systems. A participatory approach, involving tribals in the developmental process, could have created visible changes. Instead of ensuring the fruit of the developmental process, granting the right to collect and use forest produce besides regularising unauthorised lands would not sufficiently help them to move on the path of progress. These half-hearted measures contain the danger of pushing tribals further backward instead of facilitating their integration into the mainstream. Unless land alienation is either arrested or kept under check, any developmental programme aimed at empowering the tribal communities is doomed to fail.

Unfortunately, the prime focus of the 2006 Act remains the regularisation of lands held by tribals without any land deed or other documents, a measure which also benefits non-tribals. The Act is silent about the plight of the more than one-third of the tribal population who have lost their land. They have been left with no option but to go further into the forest—or to migrate elsewhere. In the Indian context, the political economy of the land and the societal power relations associated with its possession need not be emphasised. Hence, any attempt towards achieving developmental objectives should necessarily commence from restoration of the now alienated but erstwhile tribal occupied lands. Equally important is the need to ensure a participatory approach and an efficient delivery system.

That the new legislation has failed to address all these crucial issues raises serious concerns about its purpose and stated objectives. The remedial measures contained in it, besides being inadequate to tackle the problems if implemented, could turn out to be an impediment to the empowerment of the tribals and conservation as an end in itself. In other words, as Singh (2005) indicates, its main achievement will be the destruction of our forests and not the betterment of tribal living standards. Initially, the draft bill proposed to recognise the inalienable rights of the tribal communities over the forests, but in the end it has come to offer the legitimacy of induced land alienation, the root cause of the problem since the colonial period. Therefore, the 2006 Act itself appears to be politically motivated, and it is not a surprising finding that the focus on granting ownership rights to tribals was just a cloak to justify and legitimise non-tribal intervention and the disastrous consequences that are attendant on this.

Notes

1. Velayutham Saravanan, *Economic Transformation of Tribals in Tamil Nadu since the Colonial Rule 1792–1991*, unpublished PhD dissertation, Hyderabad: University of Hyderabad, 1994.
2. For details see a large number of sources: Ramachandra Guha, 'Forestry in British and Post-British India: A Historical Analysis' *Economic and Political Weekly*, 18(45–46), 1983, pp. 1886 and 1892; Ramachandra Guha and Madhav Gadgil, 'State Forestry and Social Conflict in British India' *Past and Present*, 122, 1989, pp. 148–157; Ramachandra Guha, *The Unquiet Woods: Ecological Change and Peasant Resistance in the Western Himalayas*, Delhi: Oxford University Press, 1989, pp. 29 and 185; Felix Padel, *The Sacrifice of Human Being: British Rule and the Konds of Orissa*, Delhi: Oxford University Press, 1995, pp. 28–29; Satpal Sangwan, 'Making of a Popular debate: The Indian Forester and the Emerging Agenda of State Forestry in India, 1878–1904', *Indian Economic and Social History Review*, 36(2), 1999, p. 189; Richard H. Grove, *Green Imperialism: Colonial expansion, tropical island Edens and origins of environmentalism 1600–1860*, Delhi: Oxford University Press, 1995, Chap. 8; Mahesh Rangarajan, 'Protection, Desiccation and Forest Management in the

Central Provinces 1850–1930' in Richard H Grove, Satpal Sangwan and Vinita Damodaran (eds.) *Nature and the Orient: The Environmental History of South and Southeast Asia*, Delhi: Oxford University Press, 1998, p. 589; Vandana Shiva,. *Ecology and the Politics of Survival: Conflicts over natural Resources in India*, New Delhi, 1991, p. 78; Velayutham Saravanan, 'Commercialisation of Forest, Environmental Negligence and Alienation of Tribal Rights in Madras Presidency: 1792–1882,' *Indian Economic and Social History Review*, 35(2), 1998, pp. 125–146; Velayutham Saravanan, 'Colonial Commercial Forest Policy and the Tribal Private Forests in Madras Presidency: 1792–1881', *Indian Economic and Social History Review*, 40(4), 2003, pp. 403–423; Velayutham Saravanan, 'Colonialism and Coffee Plantations: Decline of Environment and Tribals in Madras Presidency during the 19th Century', *Indian Economic and Social History Review*, 41(4), 2004, pp. 465–488; Velayutham Saravanan, 'Colonialism and Conservation: Commercialisation of Forests and decline of tribals in Madras Presidency, 1882–1947' presented at the workshop on the 'Nature in the Raj: Colonial Environments and Indigenous Knowledge' organised by Centre for World Environmental History, Sussex University and the Centre for Studies in Science Policy, Jawaharlal Nehru University, New Delhi, 2006, 18–19 August; Velayutham Saravanan, 'Tribal Welfare Measures and Colonial Policy in the Salem and Baramahal Regions of Madras Presidency during the Colonial period: 1792–1947' [Unpublished paper]; Dhirendra Datt Gangwal, 'Commercialisation of Forests, Timber Extraction and Deforestation in Uttaranchal, 1815–1947', *Conservation and Society*, 3(1), 2005, pp. 110–133; Marlene Buchy, 'British Colonial Forest Policy in South India: An Unscientific or Unadopted Policy?' in Richard H. Grove, Satpal Sangwan and Vinita Damodaran (eds.) *Nature and the Orient: The Environmental History of South and Southeast Asia*, Delhi: Oxford University Press, 1998, p. 669; N.C. Saxena, *Forest Policy in India*, New Delhi: World Wide Fund for Nature–India, and London: International Institute for Environment and Development,1999; Colyear H. Dawkins and Michael S. Philip, *Tropical Moist Forest Silviculture and Management: A History of Success and Failure*, Wallingford, UK: CAB International,1998.

3. Velayutham Saravanan, 'Economic Transformation of Tribals in Tamil Nadu: Subsistence to Commercialisation 1976–1995,' *Review of Development and Change*, 2(1), 1997, pp. 155–168; Velayutham Saravanan, 'Decline of Tribal Economy in Tamil Nadu: 1947–2000', in R.C. Rath, (ed), *Tribal Development in India: The Contemporary Debate*. New Delhi: Sage Publications, 2006, pp. 218–236; Velayutham Saravanan, 'Joint Forest Management in Andhra Pradesh: Who does the Offences?' *Asian Economic Review*, 48(2), 2006, pp. 231–238.

4. Saravanan, 'Economic Transformation of Tribals'; Saravanan, 'Economic Transformation of Tribals in Tamil Nadu: Subsistence to Commercialisation' pp. 157–168; Saravanan 'Decline of Tribal Economy in Tamil Nadu', pp. 218–236; Saravanan, 'Joint Forest Management in Andhra Pradesh', pp. 231–238.
5. Velayutham Saravanan, 'Environmental History of Tamil Nadu State, Law and Decline of Forest and Tribals, 1950–2000' *Modern Asian Studies*, 41(4), 2007, pp. 723–767; Saravanan, 'Joint Forest Management in Andhra Pradesh', pp. 231–238; Gangwal, 'Commercialisation of Forests, Timber Extraction and Deforestation'.
6. Pradip Prabhu, 'The Right to Live with Dignity', *Seminar*, 552, 2005, pp. 14–19.
7. P.V. Jayakrishnan, 'Is There a Need for this Bill?' *Seminar*, 552, 2005, pp. 23–29.
8. Ashish Kothari, 'Bungle in the Jungle' *Seminar*, 552, 2005, pp. 63–69; Prabhu, 'The Right to Live with Dignity'.
9. Apart from the sources listed in note 2 above, see for details, Indra Munshi, 'Scheduled Tribes Bill, 2005', *Economic and Political Weekly*, 40(41), 2005, p. 4407; Velayutham Saravanan, 'Tribal Revolts in India with Reference to Salem and Baramahal Districts of Madras Presidency during the late 18th Century', *Artha Vijnana*, 41(1), 1999, pp. 67–81; Velayutham Saravanan, 'Commercial Crops, Alienation of Common Property Resources and Change in Tribal Economy in the Shervaroy Hills of Madras Presidency During the Colonial Period', *Review of Development and Change*, 4(2), 1999, pp. 298–317; Velayutham Saravanan, 'Tribal Land Alienation in Madras Presidency during the Colonial Period: 1792–1947', *Review of Development and Change*, 6(1), 2000, pp. 73–104; Velayutham Velayutham Saravanan, 'Colonial Agrarian Policies in the Tribal Areas of Madras Presidency: 1872–1947', *South Asia Research*, 26(1), 2006, pp. 63–85.
10. Saravanan, 'Tribal Revolts in India with Reference to Salem and Baramahal'; Saravanan, 'Colonial Commercial Forest Policy and the Tribal Private'; Saravanan, 'Colonialism and Coffee Plantations'.
11. Bela Bhatia, 'Competing Concerns', *Economic and Political Weekly*, 40(47), 2005, pp. 4890–4893; Velayutham Saravanan, 'Thamizhaga Malaival Makkalin Nila Urimai' ['Land Rights of Tribals in Tamil Nadu'. Special Issue]. *Uzhavan Urimai*, 1999, pp. 53–57 (in Tamil); Saravanan, 'Colonial Agrarian Policies in the Tribal Areas'.
12. Saravanan, 'Colonial Agrarian Policies in the Tribal Areas'.
13. Saravanan, 'Joint Forest Management in Andhra Pradesh'.
14. Indian Council of Forestry Research and Education, *Forestry Statistics of India*. Dehra Dun: ICFRE, 2000.

15. Ibid.
16. Bhatia, 'Competing Concerns', p. 4892.
17. Saravanan, 'Economic Transformation of Tribals in Tamil Nadu'; Saravanan, 'Tribal Land Alienation in Madras Presidency'; Saravanan, 'Colonial Agrarian Policies in the Tribal Areas'.
18. Ajit Menon and Velayutham Saravanan, 'Displacement and Rehabilitation Policies: The Case of Kolli Hills Hydro–Electric Project', *Economic and Political Weekly*, 31(43), 1996, pp. 2854–2855; Velayutham Saravanan, 'Impact of Welfare Programmes on the Tribals in Tamil Nadu: 1947–1991', *Review of Development and Change*, 3(1), 1998, pp., 90–108; Velayutham Saravanan, 'Economic Transformation of Tribal Women in Tamil Nadu: 1947–1997', *Indian Journal of Labour Economics*, 42(4), 1999, pp. 699–706; Velayutham Saravanan, 'Changing Structure of Workforce Among the Social Groups in Rural Tamil Nadu', *Indian Journal of Labour Economics*, 46(4), 2003, pp. 1113–1121; Velayutham Saravanan, 'Subsistence to Dependence: Decline of Tribal Economy in Uttar Pradesh, 1947–2002' in U. Kalpagam et al. (eds.), *Labour and Poverty: Studies on Uttar Pradesh*, New Delhi: Segment Books, 2004, pp. 305–21; Saravanan, 'Joint Forest Management in Andhra Pradesh'.
19. Government of India, *National Forest Policy*, New Delhi: Ministry of Environment and Forests, 1988.
20. Saravanan, 'Impact of Welfare Programmes on the Tribals'.
21. Apparently there are six policies: (FP1) Regularisation of Encroachment; (FP2) Review of Disputed Claims over Forest Land; (FP3) Regularisation of *pattas* and leases; (FP4) Elimination of Intermediaries and Payment of Fair Wages to the Labourers on Forestry Works; (FP5) Conversion of Forest Villages into Revenue Villages and Settlement of Other Old Habitations; (FP6) Payment of Compensation for Loss of Life and Property Due to Predation/Depredation by Wild Animals.
22. Kothari, 'Bungle in the Jungle'.
23. Saravanan, 'Joint Forest Management in Andhra Pradesh'; Saravanan, 'Environmental History of Tamil Nadu'.
24. Census of India 2001. New Delhi: Office of the Registrar General and Census Commissioner.
25. Ibid.
26. Velayutham, 'Subsistence to Dependence'; Velayutham Saravanan, 'Globalisation and Non-Farm Employment in Andhra Pradesh'. Paper presented at the Conference on 'Rural Employment with Focus on the Non–farm Sector', held at the National Institute of Rural Development (NIRD), Hyderabad, 2003.

27. Saravanan, 'Tribal Land Alienation in Madras Presidency'; Saravanan 'Decline of Tribal Economy in Tamil Nadu'; Saravanan, 'Joint Forest Management in Andhra Pradesh'.
28. Velayutham Saravanan, 'Tamil Nadu: An Ecological Tragedy'. *Frontier*, 33(3), 2000, pp. 4–6; Saravanan, 'Environmental History of Tamil Nadu'.
29. E. Karuppaiyan, 'Alienation of Tribal Lands in Tamil Nadu – Panel Data Analysis', *Economic and Political Weekly*, 35(37), 2000, pp. 3343–350.
30. Ibid.
31. Debi Goenka, 'A Non-Solution', *Seminar*, 552, 2005, pp. 35–36.
32. Mihir Shah, 'First You Push Them in, Then You Throw Them Out', *Economic and Political Weekly*, 40(47), 2005, p. 4897.
33. Saravanan, 'Joint Forest Management in Andhra Pradesh'.
34. Government of India, *Report of the Task Group on Development of Scheduled Castes and Scheduled Tribes*, New Delhi: Planning Commission, 2005, p. 89.
35. Tavleen Singh, 'Lies about Tribal Rights', *Seminar*, 552, 2005, p. 38.
36. Mahesh Rangarajan, 'Tribal Bill: Fire in the Forest', *Economic and Political Weekly*, 40(47), 2005, p. 4888.
37. Singh, 'Lies about Tribal Rights'.
38. http://presidentofindia.nic.in/scripts/palatest1.jsp?id=7. [Accessed 4 December 2007].
39. Government of India, *The Scheduled Tribes (Recognition of Forest Rights) Bill, 2005*. New Delhi: Ministry of Tribal Affairs, 2005.
40. For details see Munshi (2005), and various contributions in Economic and Political Weekly, 40(47), namely Bhatia (2005), Krishnaswamy (2005), Rangarajan (2005) and Shah (2005). See also several contributions in issue 552 of Seminar (August 2005), namely Dang (2005), Goenka (2005), Jayakrishnan (2005), Kothari (2005), Mohanty (2005), Prabhu (2005) and Singh (2005).
41. Munshi, 'Scheduled Tribes Bill, 2005', p. 4406.
42. Kothari, 'Bungle in the Jungle'.
43. Himraj Dang, 'The End of Conservation', *Seminar*, 552, 2005, pp. 50–54; Munshi, 'Scheduled Tribes Bill, 2005', p. 4406; Lok Sabha Secretariat, *Joint Committee on the Scheduled Tribes (Recognition of Forest Rights) Bill, 2005*. New Delhi: Lok Sabha Secretariat, 2006.
44. Dang, 'The End of Conservation'.
45. Madhuri Krishnaswamy, 'One Step Forward, Two Steps Back', *Economic and Political Weekly*, 40(47), 2005, p. 4900.
46. Bhatia, 'Competing Concerns', p. 4891; Munshi, 'Scheduled Tribes Bill, 2005', p. 4407; Krishnaswamy, 'One Step Forward, Two Steps Back', p. 4899.
47. Singh, 'Lies about Tribal Rights'.

48. Saravanan, 'Tribal Land Alienation in Madras Presidency'; Saravanan, 'Colonialism and Coffee Plantations'; Saravanan, 'Colonial Agrarian Policies in the Tribal Areas'.
49. Indian Council of Forestry Research and Education, *Forestry Statistics of India*. Dehra Dun: ICFRE, 1996.
50. Indian Council of Forestry Research and Education, 2000.
51. Saravanan, 'Colonial Commercial Forest Policy and the Tribal Private Forests'; Saravanan, 'Colonial Agrarian Policies in the Tribal Areas'.
52. Saravanan, 'Tribal Land Alienation in Madras Presidency'; Saravanan, 'Colonialism and Coffee Plantations'.
53. J.J. Roy Burman, 'The Tribal Bill: A Rejoinder', *Economic and Political Weekly*, 40(52), 2005, p. 5515.
54. Bhatia, 'Competing Concerns', p. 4892.
55. Velayutham Saravanan, 'Scheduled Tribes Bill, 2005: Why Forest Rights and Not Land?' *Mainstream*, 43(3), 2005, pp. 22–23.
56. Bhatia, 'Competing Concerns', p. 4892; Saravanan, 'Scheduled Tribes Bill, 2005'.
57. Velayutham Saravanan, 'Terrorising Tribals: Nexus between Police and Forest Department', *Economic and Political Weekly*, 32(29), 1997, pp. 1789–1790.
58. Krishnaswamy, 'One Step Forward, Two Steps Back', p. 4901.

CHAPTER 7

Conclusion

The main objective of this book is to provide a comprehensive picture of the nature and trends in India's environmental history encompassing the colonial and post-colonial period, from the late eighteenth to the early twenty-first century (1792–2007).

Until the intervention of colonial powers, the forest-dwelling communities used the forests for their livelihood without any commercial motive. Owing to the low density of population as well as their limited needs, natural resources in the forests were not disturbed much. The interaction between forest-dwellers and the plains people was also very limited. In other words, there was no scope for the commercialisation of forest resources because of the inhabitants' isolation and subsistence livelihood. Hence, one can see forest resources in abundance at the time of colonial intervention.

After the advent of the British, not only were forest resources systematically exploited but the establishment of plantations, through the denudation of vast tracts of pristine forest cover, were also encouraged in different parts of India. No wonder that this project carried out on a very large scale resulted not only in the decline of forest resources but also deprived the forest-dwelling communities of their livelihood and their economic status. In other words, the owners of the forests became dependent on the forest encroachers. The systematic plunder of forest resources by the pre-Forest Act colonial regime destroyed the livelihood of forest-dwelling communities and tore asunder their traditional socio-economic and cultural system.

© The Author(s) 2018

V. Saravanan, *Environmental History and Tribals in Modern India*,
https://doi.org/10.1007/978-981-10-8052-4_7

The British colonial forest policy had failed to accord due importance to the environment, since it was concerned with the commercialisation of forest resources in India during the pre-Forest Act era.

The introduction of reserve forests was expected to ensure environmental protection. However, whether or not the post-Forest Act era witnessed greater environmental protection needs to be scrutinised. It appears that these conservation measures actually restricted the forest-dwelling communities from using forest resources, while the colonial administration continued to systematically extract these resources for the rest of the colonial period. The series of restrictions imposed on the forest-dwelling communities left them with no option but to depend upon the mercy of the Forest Department. In other words, the rights of the forest-dwelling communities and their movements were not only curbed but they were reduced to the status of culprits. To be precise, in its quest for the continued looting of forest wealth, the post-Forest Act colonial regime neither protected the environment nor ensured the welfare of the forest-dwelling communities. Consequently, the living conditions of the forest-dwelling communities deteriorated and the occupational status of a vast majority of them changed to that of labourers, leaving them open to harassment by officials.

Post-colonial policy was no different. Unfortunately, what were pursued were the same ends: the commercialisation of forests and the neglect of the legitimate concerns of the forest-dwelling communities until the last quarter of the twentieth century. The policy framework of post-colonial governments with regard to forests had inherent flaws and contradictions owing to the absence of vision to address environmental and tribal concerns together. On the one hand, the state continued to encroach upon forest land for various development purposes, and on the other, it offered encouragement to the denudation and plunder of forest resources for commercial gain. While restricting access to forests and the collection of forest produce by the tribals, the state has either remained a mute spectator to the loot of timber and other resources such as minerals or stood apparently ineffective in controlling the contractors/politicians/mafia nexus. Deforestation and denudation caused by the tribals has meant that curbs have been put on them, but the illicit felling and smuggling of forest wealth, despite stringent laws, have raised no alarms, and various parties have escaped accountability. Another pertinent issue is the failure of the state to provide infrastructure facilities to the tribal areas. The state has either overlooked the systemic failures of governance or failed to impose restrictions on non-tribal intruders into tribal areas. In short, in the post-colonial period there was no paradigmatic shift in the perspective of the state with regard to hill areas and the tribals. Instead, the pace of disintegration of the traditional forest-oriented tribal economy was accentuated. The policies pursued during the first three decades of the post-independence period neither protected the environment nor promoted tribal welfare, but only paved the way for the non-tribals to destroy the environment and the tribal economy.

During the 1980s, development measures were initiated among the tribals but failed to deliver any positive economic transformation for them. Government policies provided the requisite thrust for improving infrastructural facilities in tribal areas, particularly during the last two and a half decades of twentieth century. However, they have made very little impact on socio-economic development, primarily owing to ineffective delivery systems. A participatory approach, involving tribals in the developmental process, could have created visible changes. Nothing other than ensuring the fruit of the developmental programmes, granting the right to collect and use forest produce besides regularising unauthorised lands, would sufficiently help them to move on to the path of progress. As such, these half-hearted measures help to push tribals further backward instead of facilitating their integration into the mainstream.

Tribal development policies, which gave an impetus to infrastructural facilities—roads, communications, schools, hospitals, credit and so on—made an attempt to integrate the economy of the tribal communities into mainstream society. The Tribal Sub-Plan, of the last two decades of the twentieth century in particular, was aimed at accomplishing this agenda. However, these measures which were intended to spur economic development have somehow failed in many parts of the country. The various arms of the state apparatus have not only failed to deliver prompt services to improve tribals' living standards but have also exploited them economically and physically. The gruesome Vachathi police action is one of the worst state-sponsored atrocities on tribal people during the post-independence period. While failing to arrest the timber mafia who enjoy political clout, the law enforcement authorities went on a rampage in that tribal village, resulting in a militant subaltern protest against the state apparatus.

Recent policy initiatives also betray a lack of concern for the environment and tribal development. Admittedly development has been broadly accepted to mean a process of improving the status and living conditions of people, while conservation denotes the protection of natural resources. The new legislation might facilitate tribals' use of forest resources for their subsistence and guarantee them ownership rights. At the same time, it has to be borne in mind that the 2006 Act remains silent on addressing tribal land alienation, the restoration of lands from non-tribals that were occupied prior to the cut-off date. Entry of non-tribals into the hills preceded the establishment of the Forest Department in the early nineteenth century. Even after the introduction of the reserve forests during the late nineteenth century, a very large proportion of tribal lands was alienated by non-tribals. Consequently, about 37% of the tribal population was deprived of its land and by 2001 had become agricultural labourers. No one can possibly argue that tribal households didn't possess forest lands prior to the intervention of the Forest Department or the advent of non-tribal settlements. Every household occupied the land that was required for its sustenance and survival. While non-tribals garnered most of the fertile and cultivated lands, the tribals were forced to go deeper into the forests or become dependants. Hence, restoring alienated land is a more viable and sustainable option than just giving tribal people ownership rights over forest lands. Unless land alienation is either arrested or kept under check, any developmental programme aimed at empowering the tribal communities is doomed to fail.

Unfortunately, the prime focus of the 2006 Act remains the regularisation of lands held by tribals without any land deed or other documents, a measure that also benefits non-tribals. The Act is silent on the plight of the more than one-third of the tribal population who have lost their land over the period. They have been left with no option but to go further into the forest—or to migrate elsewhere.

That the new legislation has failed to address all these crucial issues raises serious concerns about its purpose and stated objectives. The remedial measures contained in it, besides being inadequate to tackle the problems if implemented, could turn out to be an impediment for the empowerment of the tribals and conservation. In other words, as Singh (2005) indicates, its main achievement will be the destruction of our forests and not the betterment of tribals' living standards. Initially, the draft bill proposed to recognise the inalienable rights of the tribal communities over the forests, but in the end it came to offer legitimacy on induced land alienation, the root cause of the problem since the colonial period. Hence, the 2006 Act itself appears to be politically motivated, and it is not a surprising finding that the focus on granting ownership rights to tribals was just a cloak to justify and legitimise non-tribal intervention and the attendant disastrous consequences.

Since the early nineteenth century, forest resources were exploited for commercial purposes by the colonial government. No constructive measures were initiated to conserve forests and wildlife resources, and it was only the belated realisation that the resources were becoming exhausted that forced the initiation of gestures towards conservation. One can see the same trend even in the post-Forest Act period, when forests were demarcated, surveys undertaken, working plans prepared and fire protection measures initiated. But for what purpose these measures were initiated? It must be remembered that they were introduced to ensure the sustainability of forest resources for commercial viability. The Forest Department has imposed a whole range of restrictions on tribals and other forest users since the late nineteenth century. Now (2007) this was recognised by the government. While introducing the Scheduled Tribes (Recognition of Forest Rights) Bill 2005, it was accepted that 'rights on ancestral lands and their habitat were not adequately recognised in the consolidation of state forests during the colonial period as well as in independent India resulting in historical injustice to the forest dwelling Scheduled Tribes who are integral to the very survival and sustainability of the forest ecosystems'. Hence, one can argue that the conservation measures of the colonial government in India were designed not to preserve forest resources but only to meet their commercial interests. Post-colonial forest policies also continued to encourage commercialisation and were still less concerned with tribal development. If there were any initiatives at all, they were by and large dictated by the demands of populist vote-bank politics.

Glossary

amani Government officers' collection revenue directly from cultivators upon the removal or suspension of an intermediate claimant.

annas The lowest money measurement. Twelve *pies* are equal to one *anna*, 16 *annas* or 192 *pies* equal to 1 Company rupee.

avarai *Cassia auriculata*.

beedi Indian cigarette filled with tobacco flake and wrapped in *tendu* leaves.

candies A dry grain measurement. One *candies* is equal to 40 *bullas* or *puddies* or 7,248 cubic inches.

cash It was a lowest money measurement. Eighty *cash* is equal to 1 *fanam* and 45 *fanams* are equal to 1 *Star Pagoda*.

Chetties A business community.

cholam *Sorghum vulgare*.

fanams A money measurement, 80 *cash* equal to one *fanam*; 45 *fanams* equal to 1 *Star Pagoda*.

guru Spiritual leader.

hartal Mass protest.

jagir A tenure common under Muslim government, in which the public revenues of a given tract of land were made over to a servant of the state, together with the power requisite to enable him to collect and appropriate such revenue and administer the general government of the district.

© The Author(s) 2018

V. Saravanan, *Environmental History and Tribals in Modern India*,
https://doi.org/10.1007/978-981-10-8052-4

kambu Pearl millet.
kodallay Groundnut, Arachis hypogea.
kothukadu The lands that are cleared, cultivated for a few years then abandoned, and covered with a fresh growth of jungle. In other words, land in the hill areas capable of cultivation after long intervals with particular kinds of grain sown in holes dug with a spade.
Malaiyalis A tribe in Tamil Nadu.
mochaikottay Field bean, Volichos lab.
mounds A measure of weight that prevailed in Madras Presidency. One *mound* is equal to 25 lbs.
nullahs A watercourse, riverbed, or ravine.
panivaragu Proso millet.
Pariahs A Scheduled Caste under the Indian Constitution.
pattah Land deed or ownership rights of land.
ragy **or** *kevaroo* Finger millet.
Sowcars Moneylenders.
Star Pagodas One *Star Pagoda* was equal to 45 *fanams* or 3600 *cash*. One *Star Pagoda* is equal to 3½ Company rupees.
tatapyru A kind of pulse.
tatties A matted screen.
tendu *Diospyros meanoxylon.*
thinay Foxtail millet.
thovarai Redgram, Cajanus indicus.
varagu Kodo millet.

Bibliography

Administration Report of the Forest Department of the Madras Presidency (various years).

Administrative Report of the Madras Presidency (various years).

Adolphus, W.H.W. *Consolidated Working Plan for the Central Salem Division*. Madras: Government Press, 1932.

Arbuthnot, A.J. *Sir Thomas Munro: Selections from Minutes and other Official Writings*, 2 volumes. London: C. Kegan Paul & Co., 1881.

Arnold, David, and Guha, Ramachandra. 'Introduction: Themes and Issues in Environmental History of South Asia' in David Arnold, and Ramachandra Guha (eds), *Nature, Culture, Imperialism: Essays on the Environmental History of South Asia*. Delhi: Oxford University Press, 1995, pp. 1–20.

Baker, David. 'State polity, the market economy, and Tribal decline: The central provinces, 1861–1920'. *Indian Economic and Social History Review*, 1991, 28(4): 341–370.

Bandopadhyay, Arun. 'Three issues from a CPR management: Village forestry in post-colonial South Asia' in B.B. Chaudhuri and Arun Bandopadhyay (eds), *Tribes, Forest and Social Formation in Indian History*. New Delhi: Manohar, 2004, pp. 209–224.

Basu, Kaushik. *The Retreat of Democracy: Globalization, Economics and the India*. Delhi: Permanent Black, 2007.

Beinart, William, and Hughes, Lotte. *Environment and Empire*. Oxford: Oxford University Press, 2007.

Bhadran, C.A.R. 'Forests of the Madras State' in *Forest Centenary of Forest Administration in Madras State*. Madras: Madras Forest Department, 1959, pp. x–xi.

© The Author(s) 2018

V. Saravanan, *Environmental History and Tribals in Modern India*,
https://doi.org/10.1007/978-981-10-8052-4

Bhamathi, B. *Report of the on-the spot enquiry into the atrocities on sch. Tribes of Vachathi village, Harur Taluk, Dharmapuri District by Forest officials and Police.* Madras, Director for SCs and STs (dated 19–8–1992), 1992a.

Bhamathi, B. *Report on the prevailing conditions in Vachathi village.* Madras, Director for SCs and STs (dated 22–12–1992), 1992b.

Bhatia, Bela. 'Competing Concerns'. *Economic and Political Weekly*, 2005, 40(47): 4890–4893.

Bhattacharya, Neeladri. 'Pastoralists in a Colonial World' in David Arnold, and Ramachandra Guha (eds), *Nature, Culture, Imperialism: Essays on the Environmental History of South Asia.* Delhi: Oxford University Press, 1995, pp. 49–85.

Brandis, Dietrich. *Suggestions Regarding Forest Administration in the Madras Presidency.* Madras: Government Press, 1883.

Buchanan, Francis. *A Journey from Madras through the Countries of Mysore, Canara and Malabar*, 3 Vols. London: Directors of the East India Company, 1807.

Buchy, Marlene. 'British Colonial Forest Policy in South India: An Unscientific or Unadopted Policy?' in Richard H. Grove, Satpal Sangwan and Vinita Damodaran (eds), *Nature and the Orient: The Environmental History of South and Southeast Asia.* Delhi: Oxford University Press, 1998, pp. 636–673.

Census of India 1871, *Census of the Madras Presidency 1871 A Report*, Vol. I. Madras: Government Gazette Press, 1874a.

Census of India 1871, *Census Statement of population 1871, Salem District.* Madras: Government Gazette Press, 1874b.

Census of India 1891, *Taluqwar and Village Statistics, Salem District.* Madras: Government Press, 1893.

Census of India 1951, *Census Handbook, Salem District.* Madras: Government Press, 1953.

Census of India 1961, *Census Handbook, Salem District.* Madras: Government Press, 1964.

Census of India 1971, *Special Tables on Scheduled Castes and Scheduled Tribes.* Madras: Government Press, 1977.

Census of India 1981, *Special Tables for Scheduled Tribes.* Madras: Government Press.

Census of India 1991, *Primary Census Abstract for Scheduled Castes and Scheduled Tribes.*

Census of India 2001. New Delhi: Office of the Registrar General and Census Commissioner.

Chandran, Subash M.D. 'Shifting Cultivation, Sacred Groves and Conflicts in Colonial Forest Policy in Western Ghats' in Grove, Richard H., Sangwan, Satpal and Damodaran, Vinita (eds) *Nature and the Orient: The Environmental History of South and Southeast Asia.* Delhi: Oxford University Press, 1998, pp. 674–707.

Chaudhuri, K.N. 'Foreign Trade' in Tapan Raychaudhuri and Irfan Habib (eds), *The Cambridge Economic History of India*, Vol. 1. Cambridge: Cambridge University Press, 1982, pp. 382–433.

Chaudhuri, K.N. *Economic Development of India under the East India Company, 1814–1858: A selection of contemporary writings in Cambridge*. Cambridge: Cambridge University Press, 1971.

Chaudhury, B.B. 'Adivasi and Aranyaka: Reconsidering some characterizations of their polity and economy in pre-colonial and colonial India' in B.B. Chaudhuri and Arun Bandopadhyay (eds), *Tribes, Forest and Social Formation in Indian History*. New Delhi: Manohar, 2004, pp. 89–108.

Cox S. The Forest of Madras. in Somerset Playne (compiled) *South India*. London: Foreign and Colonial Compiling and Publishing & Co. 1914–1915.

Cronon, William. 'The Uses of Environmental History'. *Environmental History Review*, 1993, 17(3): 1–22.

Crosby, Alfred W. *Ecological Imperialism: The Biological Expansion of Europe, 900–1900*. Cambridge: Cambridge University Press, 1986.

Dang, Himraj. 'The End of Conservation'. *Seminar*, 2005, 552: 50–54.

Dawkins, H. Colyear, and Philip, Michael S. *Tropical Moist Forest Silviculture and Management: A History of Success and Failure*. Wallingford: CAB International, 1998.

Dhinamani July 28, 1992 (Tamil Newspaper).

Food and Agriculture Organisation. Flavours and Fragances of Plant Origin, 1995. http://www.fao.org/docrep/v5350e/V5350e08.htm.

Gadgil, Madhav, and Guha, Ramachandra. *This Fissured Land: An Ecological History of India*. Delhi: Oxford University Press, 1992.

Gadgil, Madhav. 'Towards an Ecological History of India'. *Economic and Political Weekly*, 1985, 20(45, 46 and 47): 1913.

Gangwal, Dhirendra Datt. 'Commercialisation of Forests, Timber Extraction and Deforestation in Uttaranchal, 1815–1947'. *Conservation and Society*, 2005, 3(1): 10–33.

Goenka, Debi. 'A Non-Solution'. *Seminar*, 2005, 552: 35–36.

Government of India *Annual Plan for 1989–1990*. New Delhi.

Government of India. *National Forest Policy, 1988*. New Delhi: Ministry of Environment and Forests, 1998.

Government of India. *Report of the Task Group on Development of Scheduled Castes and Scheduled Tribes*. New Delhi: Planning Commission, 2005.

Government of India. The Scheduled Tribes (Recognition of Forest Rights) Bill, 2005, No.17014/4/2005–S&M(Pt.), New Delhi: Ministry of Tribal Affairs, 2005.

Government of Madras. *Third Five-Year Plan*. Madras: Government Press, 1960.

Government of Tamil Nadu. *Ninth Five-Year Plan 1997–2002*. Chennai: Government Press, 1998.

Government of Tamil Nadu *An Economic Appraisal* (various years).
Government of Tamil Nadu Tamil Nadu Sandalwood Possession Rules, 1970.
Government of Tamil Nadu Tamil Nadu Sandalwood Transit Rules, 1967.
Government of Tamil Nadu *Tamil Nadu State Administrative Report 1989–1990*, 1993.
Government of Tamil Nadu, *Policy Note on Forest department* (various years).
Government of Tamil Nadu. *Fifth Five-Year Plan*. Madras: Government Press, 1973.
Government of Tamil Nadu. *Fourth Five-Year Plan*. Madras: Government Press, 1969.
Government of Tamil Nadu. *Seventh Five-Year Plan 1985–1990*. Madras: Government Press, 1985.
Government of Tamil Nadu. *Synoptic Statistics on Forestry in Tamil Nadu*. Madras: Office of the Chief Conservator of Forests, 1978.
Government of Tamil Nadu. Tamil Nadu Forest Act 1882.
Government of Tamil. *Eighth Five-Year Plan 1992–1997*. Madras: Government Press, 1992.
Grove, Richard H. *Green Imperialism: Colonial Expansion, Tropical Island Edens and Origins of Environmentalism 1600–1860*. Delhi: Oxford University Press,1995.
Guha, Ramachandra, and Madhav Gadgil. 'State Forestry and Social Conflict in British India'. *Past and Present*, 1989, 122: 141–177.
Guha, Ramachandra. 'Forestry and Social protest in British Kumaon, c.1893–1921' in Ranajit Guha (ed), *Subaltern Studies IV: Writings on South Asian History and Society*. Delhi: Oxford University Press, 1985, pp. 55–100.
Guha, Ramachandra. 'Writing Environmental History in India'. *Studies in History*, n.s, 1993, 9(1): 119–129.
Guha, Ramachandra. 'Forestry in British and Post-British India: A Historical Analysis'. *Economic and Political Weekly*, 1983, 18(45–46): 1886–1892.
Guha, Ramachandra. *The Unquiet Woods: Ecological Change and Peasant Resistance in the Western Himalayas*. Delhi: Oxford University Press, 1989.
Guha, Sumit. 'Claims on the commons: Political Power and Natural Resources in Pre-Colonial India'. *Indian Economic and Social History Review*, 2002, 39 (2&3): 181–196.
Guha, Sumit. 'Forest politics and Agrarian empires: The Kandesh Bhills, c.1700–1850'. *Indian Economic and Social History Review*, 1996, 33(2): 133–153.
Guha, Sumit. *Environment and Ethnicity in India, 1200–1991*. Cambridge: Cambridge University Press, 1998.
Hardiman, David. *Peasant Resistance in India 1858–1914* (ed). Delhi: Oxford University Press, 1992.
http://www.thehindu.com/2003/03/13/stories/2003031305750400.htm.

http://www.thehindu.com/2003/05/02/stories/2003050204850400.htm.

Hurd, John M. 'Railways' in Dharma Kumar and Meghnad Desai (eds), *The Cambridge Economic History of India*, Vol. 2. Cambridge: Cambridge University Press, 1983, pp. 737–761.

Indian Council of Forest Research & Education. *Forestry Statistics India*. Dehra Dun: ICFRE, 1995.

Indian Council of Forestry Research and Education. *Forestry Statistics of India*. Dehra Dun: ICFRE, 1996.

Indian Council of Forestry Research and Education. *Forestry Statistics of India*. Dehra Dun: ICFRE, 2000.

Indian Express, dated July 19, 1992 (Chennai edition).

Institute of Techno-Economic Studies. *Indebtedness of Scheduled Tribes in Tamil Nadu*. Madras: Institute of Techno-Economic Studies, 1978.

Jayakrishnan, P.V. 'Is There a Need for this Bill?' *Seminar*, 2005, 552: 23–29.

Jodha, N.S. *Life on the Edge: Sustaining Agriculture and Common Property Resources in Fragile Environment*. Delhi: Oxford University Press, 2001.

Jothimurugan, A. 'Tribal Development in Tamil Nadu' Unpublished M.Phil dissertation, Madras: University of Madras, 1985.

Jothimurugan, A. 'Socio-Economic Impact of Tribal Development Schemes in Tamil Nadu: A Comparative Studies of Kalroyan Hills of Salem and South Arcot Districts' Unpublished Ph.D dissertation, Madras: University of Madras, 1989.

Karuppaiyan, E. 'Alienation of Tribal Lands in Tamil Nadu: Panel Data Analysis'. *Economic and Political Weekly*, 2000, 35(37): 3343–3350.

Kothari, Ashish. 'Bungle in the Jungle'. *Seminar*, 2005, 552: 63–69.

Krishnaswamy, Madhuri. 'One Step Forward, Two Steps Back'. *Economic and Political Weekly*, 2005, 40(47): 4899–4901.

Kumar, Dharma. *Land and Caste in South India: Agricultural labour in Madras presidency in the nineteenth century*. Cambridge: Cambridge University Press, 1965.

Kumar, Dharma. *The Cambridge Economic History of India, Volume 2: c.1750–c.1970*. Cambridge: Cambridge University Press, 1983.

Kurien, C.T. *The Economy: An Interpretative Introduction*. New Delhi: Sage Publications, 1992.

Le Fanu, H. A *manual of the Salem District in the Presidency of Madras*, Vol. II. Madras: Government Press, 1883.

Locher, Fabien, and Quenet, Grégory. 'Environmental History: The Origins, Stakes, and Perspectives of a New Site for Research' *Revue d'histoire moderne et contemporaine*, 2009/4 No. 56–4, pp. 7–38 (Fabien Locher, Grégory Quenet, 'L'histoire environnementale: Origines, enjeux et perspectives d'un nouveau chantier'. *Revue d'histoire moderne et contemporaine* 2009/4 (No 56–4), 2009, pp. 7–38. (http://www.cairn-int.info/focus-E_RHMC_564_0007--environmental-history.htm. [Accessed 3 May 2016])

Lok Sabha Secretariat. *Joint Committee on the Scheduled Tribes (Recognition of Forest Rights) Bill, 2005*. New Delhi: Lok Sabha Secretariat, 2006.

Maclean, C.D. *Manual of the Administration of the Madras Presidency*, 3 Vols. Madras: Government Press, 1885.

McNeill, J.R. 'Observations on the Nature and Culture of Environmental History'. *History and Theory*, 2003, 42(4): 5–43.

Menon, Ajit, and Velayutham Saravanan. 'Displacement and Rehabilitation Policies: The Case of Kolli Hills Hydro-Electric Project'. *Economic and Political Weekly*, 1996 31(43): 2854–2855.

Meyer, Eric. 'Forests, Chena Cultivation, Plantations and the Colonial State in Ceylon 1840–1940' in Richard H Grove, Satpal Sangwan, and Vinita Damodaran (eds), *Nature and the Orient: The Environmental History of South and Southeast Asia*. Delhi: Oxford University Press, 1998, pp. 793–827.

Mishra, B.K., Bhanu, B.A., Reddy, K.N., Sethuraman, M., and Natarajan, T.S. *Tribes of Southern Region [A Select Bibliography]*. New Delhi: Inter-India Publications, 1968.

Mohanty, Biswajit. 'A Field View'. *Seminar*, 2005, 552: 30–34.

Morris, David Morris, and Clyde B. Dudley. 'Selected Railway Statistics for the Indian Subcontinent (India, Pakistan and Bangaladesh, 1853–1946–47'. *Artha Vijnana*, 1975, 17(3): 187–222.

Munshi, Indra. 'Scheduled Tribes Bill, 2005'. *Economic and Political Weekly*, 2005, 40(41): 4406–4409.

Padel, Felix. 'Forest Knowledge: Tribal People, their Environment and the Structure of Power' in Richard H Grove, Satpal Sangwan and Vinita Damodaran (eds) *Nature and the Orient: The Environmental History of South and Southeast Asia*. Delhi: Oxford University Press, 1998, pp. 891–917.

Padel, Felix. *The Sacrifice of Human Being: British Rule and the Konds of Orissa*. Delhi: Oxford University Press,1995.

Papers relating to the Coffee districts of the Madras Presidency. Madras: Saint George Gazette Press, 1859.

Parasher-Sen, Aloka. 'Of tribes, hunders and Barbarians: Forest dwellers in the Mauryan period' in *Studies in History* (n.s) 1998, 14(2): 173–191.

Pathak, A. *Law, Strategies, Ideologies: Legislating Forests in Colonial India*. Delhi: Oxford University Press, 2002.

People's Union for Civil Liberties, Tamil Nadu. *Report of the Fact-finding team that enquired into Police/Forestofficials atrocities on Vachathi Village, Harur Taluka, Dharmapuri District, on 20th and 21st June, 1992*, PUCL, Tamil Nadu Unit, 1992.

Polanyi, Karl. *Great Transformations*. Boston: Beacon Press, 1944.

Pouchepadass, J. 'Colonialism and environment in India: Comparative perspectives'. *Economic and Political Weekly*, 1995a, 30(33): 2059–2067.

Pouchepadass, Jacques. 'British Attitudes Towards Shifting Cultivation in Colonial South India: A Case Study of South Canara District 1800–1920' in David Arnolad and Ramachandra Guha (eds) *Nature,Culture, Imperialism: Essays on the Environmental History of South Asia*. Delhi: Oxford University Press, 1995b, pp. 123–151.

Prabhu, Pradip. 'The Right to Live with Dignity'. *Seminar*, 2005, 552: 14–19.

Pratap, Ajay. *The Hoe and the Axe: An Ethnohistory of Shifting Cultivation in Eastern India*. Delhi: Oxford University Press, 2000.

Radjn, P. *The World of Primitive Man*. New York: Henry Schuman, 1953.

Raghavaiyangar, S.S. *Memorandum on the Progress of the Madras Presidency during the last Forty Years of British Administration*. Madras: Government Press, 1893.

Rajannan, Busnagi. *Salem Cyclopedia*. Salem: Institute of Kongu Studies, 1992.

Ramanathan, C. Indian Sandalwood Trade. In TED Case Studies: Sandalwood Case, 1997. www.american.edu/projects/mandala/TED/sandalwood.htm.

Ramanathan, S. *Tribal Welfare in Salem District: The Role of Government and Voluntary Organisations*. Madras: Sangam Publishers, 1977.

Rangarajan, Mahesh. 'Environmental Histories of India: Of States, Landscapes, and Ecologies' in Edmund Burke III and Kenneth Pomeranz (eds), *The Environment and World History*. Berkeley: University of California Press, 2009, pp. 229–254.

Rangarajan, Mahesh. 'Environmental Histories of South Asia: A Review Essay'. *Environment and History*, 1996a, 2(2): 129–143.

Rangarajan, Mahesh. *Fencing the Forest: Conservation and ecological change in India's central province 1860–1914*. Delhi: Oxford University Press, 1996b.

Rangarajan, Mahesh. 'Protection, Desiccation and Forest Management in the Central Provinces 1850–1930' in Richard H. Grove, Satpal Sangwan, and Vinita Damodaran (eds), *Nature and the Orient: The Environmental History of South and Southeast Asia*. Delhi: Oxford University Press, 1998, pp. 575–595.

Rangarajan, Mahesh. 'Tribal Bill: Fire in the Forest'. *Economic and Political Weekly*, 2005, 40(47): 4888–4890.

Rao, B. Janarthan. *Land Alienation in Tribal Areas*. Warangal: Kakatiya School of Public Administration Publication, 1987.

Reddy, Prakash G. *Politics of Tribal Exploitation*. New Delhi, 1987.

Richards, F.J. *Madras District Gazetteers, Salem District*, Vol. I. Part I, Madras: Government Press, 1918.

Rovers, Stephen R. 'On the Contribution of Environmental History to Current Debate and Policy'. *Environment and History*, 2000, 6(2): 131–150.

Roy Burman, J.J. 'The Tribal Bill: A Rejoinder'. *Economic and Political Weekly*, 2005, 40(52): 5514–5515.

Sahlins, Marshall. 'Tribal Economics' in George Dalton (ed), *Economic Development and Social Change*. New York: Natural History Press, 1971.

Sangwan, Satpal. 'Making of a Popular Debate: The Indian Forester and the Emerging Agenda of State Forestry in India, 1878–1904'. *Indian Economic and Social History Review*, 1999, 36(2): 187–237.

Sarada, Raj A. *Economic Conditions in the Madras Presidency 1800–1850*. Madras: University of Madras, 1941.

Saravanan, Velayutham. 'Agrarian policies in the tribal areas of Madras Presidency during the pre-survey and settlement period, 1792–1872'. *Indian Journal of Agricultural Economics*, 2010a, 65(2): 261–276.

Saravanan, Velayutham. 'Subalterns vs State Institutions: Politicians, State, Forest, Law and Atrocities on Tribals in Tamil Nadu, 1990–2000'. *The International Journal of Human Rights*, 2010b, 15(6): 948–968.

Saravanan, Velayutham. 'Economic Exploitation of Forest Resources in South India during Pre-Forest Act Colonial Era, 1793–1882'. *International Forestry Review: The International Journal of Forest Science and Policy*, 2008, 10(1): 65–73.

Saravanan, Velayutham. 'Political Economy of Recognition of Forest Rights Act, 2006: Conflict between Environment and Tribal Development'. *South Asia Research*, 2009, 29(3): 199–221.

Saravanan, Velayutham. 'Changing Structure of Workforce Among the Social Groups in Rural Tamil Nadu'. *Indian Journal of Labour Economics*, 2003a, 46(4): 1113–1121.

Saravanan, Velayutham. 'Colonial Commercial Forest Policy and the Tribal Private Forests in Madras Presidency: 1792–1881'. *Indian Economic and Social History Review*, 2003b, 40(4): 403–423.

Saravanan, Velayutham. 'Globalisation and Non-Farm Employment in Andhra Pradesh'. Paper presented at the Conference on 'Rural Employment with Focus on the Non-Farm Sector', held at the National Institute of Rural Development (NIRD), Hyderabad, 2003c.

Saravanan, Velayutham. 'Colonialism and Coffee Plantations: Decline of Environment and Tribals in Madras Presidency during the 19th Century'. *Indian Economic and Social History Review*, 2004a, 41(4): 65–88.

Saravanan, Velayutham. 'Subsistence to Dependence: Decline of Tribal Economy in Uttar Pradesh, 1947–2002' in U. Kalpagam et al. (eds.), *Labour and Poverty: Studies on Uttar Pradesh*. New Delhi: Segment Books, 2004b, pp. 305–321.

Saravanan, Velayutham. 'Colonial Agrarian Policies in the Tribal Areas of Madras Presidency: 1872–1947'. *South Asia Research*, 2006a, 26(1): 63–85.

Saravanan, Velayutham. 'Colonialism and Conservation: Commercialisation of Forests and Decline of Tribals in Madras Presidency, 1882–1947. Paper presented at the workshop on the 'Nature in the Raj: Colonial Environments and Indigenous Knowledge', organised by Centre for World Environmental History, Sussex University and the Centre for Studies in Science Policy, Jawaharlal Nehru University, New Delhi, 2006b.

Saravanan, Velayutham. 'Decline of Tribal Economy in Tamil Nadu: 1947–2000' in R.C. Rath (ed), *Tribal Development in India: The Contemporary Debate*. New Delhi: Sage Publications, 2006c, pp. 218–236.

Saravanan, Velayutham. 'Economic Reforms and Tribals in Tamil Nadu: 1991–2003'. *Social Action*, 2006d, 56(1): 25–39.

Saravanan, Velayutham. 'Joint Forest Management in Andhra Pradesh: Who Does the Offences?'. *Asian Economic Review*, 2006e, 48(2): 231–238.

Saravanan, Velayutham. 'Commercial Crops, Alienation of Common Property Resources and Change in Tribal Economy in the Shervaroy Hills of Madras Presidency During the Colonial Period'. *Review of Development and Change*, 1999a, 4(2): 298–317.

Saravanan, Velayutham. 'Economic Transformation of Tribal Women in Tamil Nadu: 1947–1997'. *Indian Journal of Labour Economics*, 1999b, 42(4): 699–706.

Saravanan, Velayutham. 'Thamizhaga Malaival Makkalin Nila Urimai' ['Land Rights of Tribals in Tamil Nadu' Special Issue], *Uzhavan Urimai*: 1999c, pp. 53–57 (in Tamil).

Saravanan, Velayutham. 'Tribal Revolts in India with Reference to Salem and Baramahal Districts of Madras Presidency during the late 18th Century'. *Artha Vijnana*, 1999d, 41(1): 67–81.

Saravanan, Velayutham. 'Economic Transformation of Tribals in Tamil Nadu: Subsistence to Commercialisation 1976–1995'. *Review of Development and Change*, 1997a, 2(1): 157–168.

Saravanan, Velayutham. 'Terrorising Tribals: Nexus between Police and Forest Department'. *Economic and Political Weekly*, 1997b, 32(29): 1789–1790.

Saravanan, Velayutham. 'Commercialisation of Forest, Environmental Negligence and Alienation of Tribal Rights in Madras Presidency: 1792–1882'. *Indian Economic and Social History Review*, 1998a, 35(2): 125–146.

Saravanan, Velayutham. 'Economic Transformation of Tribals in Tamil Nadu: Causes of Migration and its Employment Pattern, 1792–1991'. *Indian Journal of Labour Economics*, 1998b, 41(4): 695–705.

Saravanan, Velayutham. 'Impact of Welfare Programmes on the Tribals in Tamil Nadu: 1947–1991'. *Review of Development and Change*, 1998c, 3(1): 90–108.

Saravanan, Velayutham. 'Environmental History of Tamil Nadu State, Law and Decline of Forest and Tribals, 1950–2000'. *Modern Asian Studies*, 2007, 41(4): 723–767.

Saravanan, Velayutham. 'Scheduled Tribes Bill, 2005: Why Forest Rights and Not Land?' *Mainstream*, 2005, 43(3): 2–3.

Saravanan, Velayutham. 'Tamil Nadu: An Ecological Tragedy'. *Frontier*, 2000a, 33(3): 4–6.

Saravanan, Velayutham. 'Tribal Land Alienation in Madras Presidency During the Colonial Period: 1792–1947'. *Review of Development and Change*, 2000b, 6(1): 73–104.

Saravanan, Velayutham. *Colonialism, Environment and Tribals in South India, 1792–1947.* London: Routledge, 2017.

Saravanan, Velayutham. *Economic Transformation of Tribalsl in Tamil Nadu since the Colonial Rule 1792–1991.* Unpublished PhD Dissertation, Hyderabad: University of Hyderabad, 1994.

Saravanan, Velayutham. *The Socio-Economic Conditions of Scheduled Tribes: With Special Reference to Malaiyalis of Peddanaicken Palayam Block in Salem District.* Unpublished M.Phil., dissertation, Madras: University of Madras, 1989.

Saravanan, Velayutham. *Working of Nutritious Noon Meal Scheme in Kalrayan Hills: With Special Reference to Peddanaicken Palayam Block in Salem District.* Unpublished M.A., dissertation, Tiruchirappalli: Bharathidasan University, 1987.

Saxena, N.C. *Forest Policy in India.* New Delhi and London: World Wide Fund for Nature-India and International Institute for Environment and Development, 1999.

Season and Crop report of the Madras Presidency, Madras (various years).

Shah, Mihir. 'First You Push Them in, Then You Throw Them Out'. *Economic and Political Weekly*, 2005, 40(47): 4895–4899.

Shiva, Vandana. *Ecology and the Politics of Survival: Conflicts over natural Resources in India.* New Delhi: Sage Publications, 1996.

Shortt, John. *Hill Ranges of Southern India* (ed). Madras: Higginbotham and Co., 1870.

Singh, Chetan. 'Forests, Pastoralists and Agrarian society in Mughal India' in David Arnold and Ramachandra Guha (eds), *Nature, Culture, Imperialism: Essays on the Environmental History of South Asia.* Delhi: Oxford University Press, 1995.

Singh, K.S. Re-thinking forest, forest dwellers and ecological history' in B.B. Chaudhuri and Arun Bandopadhyay (eds), *Tribes, Forest and Social Formation in Indian History.* New Delhi: Manohar, 2004, pp. 39–49.

Singh, Tavleen. 'Lies about Tribal Rights'. *Seminar*, 2005, 552: 37–40.

Sivaramakrishnan, K. 'Science, Environment and Empire History: Comparative Perspectives from Forests in Colonial India'. *Environment and History*, 2008, 14(1): 41–65.

Sivaramakrishnan, K. 'Forests and the environmental history of modern India'. *Journal of Peasant Studies*, 2009, 36(2): 299–324.

Skaria, Ajay. 'Timber Conservancy, Desiccationism and Scientific Forestry: The Dangs 1840s–1920s' in Richard H Grove, Satpal Sangwan and Vinita Damodaran (eds), *Nature and the Orient: The Environmental History of South and Southeast Asia.* Delhi: Oxford University Press, 1998, pp. 596–635.

Statistical abstract relating to British India, London (various years).

Stebbing, E.P. *The Forests of India*, Vol. I. London: John Lane The Bodley Head Limited, 1922.

Stebbing, E.P. *The Forests of India*, Vol. IV. London: Oxford University Press, 1962.

Subrahmanyam, Sanjay. *Political Economy of Commerce Southern India 1500–1650.* Cambridge: Cambridge University Press, 1990.

Swami, Vandana. 'Environmental History and British Colonialism in India: A Prime Political Agenda'. *CR: The New Centennial Review*, 2003, 3(3): 113–130.

Thangam, E.S. 'Forest Products of Madras State' in *Forest Centenary of Forest Administration in Madras State.* Madras: Madras Forest Department, 1959, pp. 17–21.

Thangaprakasam, M. *Socio-economic Development of the Tribals in the Jawadhi-Elagiri Hills 1900–1984.* Unpublished M.Phil dissertation, Madras: University of Madras, 1986.

The Hindu dated 13 March 2003, Chennai.

The Hindu dated 2 May 2003, Chennai.

Thurnwald, Richard. *Economics in Primitive Communities.* London: Oxford University Press, 1932.

Viljoen, Stephen. *The Economic of Primitive People.* New York: Staples Press Ltd., 1936.

World Agricultural Census 1979–1980. Tamil Nadu, Part II, *State and Regional Tables.* Madras: Commissioner of Agricultural Census, 1989.

World Commission on Environment and Development. *From One Earth to One World: An Overview.* Oxford: Oxford University Press, 1987.

Writ Petition No.14949 of 1992 Report of CBI on the Investigation in case No. RC.3/5/95 of CBI, SCB, Madras in Compliance with the Orders Passed by the Hon'ble High Court of Madras in Writ Petition No. 14949/92 Filed by A. Nallasivam, Member of Parliament, 1992.

Index[1]

[1] Note: Page number followed by "n" refer to notes.

© The Author(s) 2018

V. Saravanan, *Environmental History and Tribals in Modern India*,
https://doi.org/10.1007/978-981-10-8052-4

FSC
www.fsc.org
MIX
Papier | Fördert
gute Waldnutzung
FSC® C083411

Kolibri 360 GmbH
Ferdinand-Jühlke-Straße 7
99095 Erfurt, Deutschland
produktsicherheit@kolibri360.de